V&R

Religion, Theologie und Naturwissenschaft/
Religion, Theology, and Natural Science

Herausgegeben von
Antje Jackelén, Gebhard Löhr, Ted Peters
und Nicolaas A. Rupke

Band 9

Vandenhoeck & Ruprecht

Anne L.C. Runehov

Sacred or Neural?

The Potential of Neuroscience to Explain Religious Experience

Vandenhoeck & Ruprecht

Mit 5 Abbildungen

Bibliografische Information der Deutschen Nationalbibliothek

Die Deutsche Nationalbibliothek verzeichnet diese Publikation in der Deutschen Nationalbibliografie; detaillierte bibliografische Daten sind im Internet über http://dnb.d-nb.de abrufbar.

ISBN 978-3-525-56980-1

Umschlagabbildung: So far, so close © Erika Bengtsdotter

Printed in Germany.
Druck und Bindung: ⊕ Hubert & Co, Göttingen.

Gedruckt auf alterungsbeständigem Papier.

Table of Contents

List of Figures

Acknowledgements

I am privileged and very grateful to a host of people who have helped me in various ways in the course of the writing of this work.

First and foremost, I owe my wholehearted gratitude to Professor Dr. Mikael Stenmark for his intellectual, and inspiring way to engage in the project. I also owe wholehearted gratitude to Professor Dr. Wentzel van Huyssteen for a thorough reading and careful commenting on my work. I also want to express my cordial thanks to Professor Dr. Niels Henrik Gregersen for the opportunity to discuss the central parts of this project. His expertise in the research field of science and theology and the thoughts he shared with me improved my work markedly. I am also in dept to Associate Professor Em., M.D. Mogens Dalby who commented on the neuroscientific part of my work, which resulted in a deepened confidence in what I was doing and a deepened understanding of neuroscience.

My sincere gratitude is also due to Professor Dr. Sven Erik Brodd, Professor Dr. Carl-Reinhold Bråkenhielm, Associate Professor Dr. Ulf Görman, Professor Dr. Eberhard Herrmann, Dr. Åsa Löwén, Professor Dr. Troels Nørager, Professor Dr. Palmyre Oomen, Professor Dr. Catharina Stenqvist and Professor Dr. Kees van Kooten Niekerk, for reading and commenting on parts of my work. In this connection I also want to thank the participants of the Research Seminar for Philosophy of Religion at the Uppsala University and the Research Seminar for Systematic Theology at the Århus University for helpful discussions.

I am also indebted to Professor Dr. Carol Albright, Professor Dr. Hans Ferdinand Angel, Associate Professor Dr., M.D. Nina Azari, Professor Dr. John H. Brooke, Professor M.D. Stefan Einhorn, Dr. Patrik Fridlund, Associate Professor Dr. Pehr Granqvist, Professor Dr. Jan-Olav Henriksen, Bishop Dr. Antje Jackelén, Professor Dr. Palmyre Oomen, Professor Em. Dr. Dr. Helmut Reich and Dr. Chris Wiltsher for thoughtful responses to papers, presentations and parts of my work.

Several people have contributed in a specific way to the quality and lay out of my book. For remarkably improving the language I owe Joyce Brooks many thanks. I am honoured and grateful to painter Erika Bengtsdotter for the permission to decorate the cover of this book with one of her beautiful pieces of art. All my informants were undoubtedly of great value.

Needless to say I am also very thankful for the love and understanding with which my husband Hans Runehov, my son Thierry Galle and his wonderful girlfriend Wendy Vanden Bossche, have surrounded me. Not only have Hans and Thierry been of immense help to me with their expertise on computers, but they have also contributed with philosophical ideas that were fruitful for my project. As a matter of fact, it was my husband who encouraged me to investigate neuroscientific views on religious experiences. Furthermore, nobody can function in an optimal way without a strong backbone. That backbone is my family. From the south of Sweden, Belgium and Russia, they have provided me with love and kept me continually in their prayers. I am deeply grateful.

Last but not least, I want to thank the head of department of Systematic Theology at Copenhagen University, Professor Dr. Bengt Flemming Nielsen for the generousity and opportunity to continue my research.

Copenhagen, 16th June 2006 Anne L.C. Runehov

Aan mijn lieve Mama
Met dankbaarheid en liefde

Ik weet haast niets meer
van alles wat ik eens heb willen zeggen.
Ik will haast niets meer zeggen.
Alleen iets van het licht.

Hans Andreus

1 Introduction

One of my informants, whom I called Cindy, described how she remembers her first religious experience, as follows:

> I remember the first time [that I had a religious experience]. I felt filled with the Holy Spirit. I was attending mass together with many other people. At a certain moment, the preacher said that we could, there and then, be baptised in the Holy Spirit and encouragingly welcomed us to come forward to the altar. However, during the mass I had been quite irritated with the preacher's way of preaching and by his conduct, which rendered a critical attitude towards him. I did not want to be commanded to go to the altar. I only wanted to go voluntarily. Therefore, I remained seated in the church bench and watched others go towards the altar. I tried to forget the preacher by concentrating on the big cross at the front of the church and by being silent before the Lord. It was then that it happened. I was smitten by an indescribable feeling of happiness. Suddenly, my soul was transformed into a source of bubbling murmuring water. The source burst through my eyes and I cried as never before. I opened my mouth and out came a language that I did not understand. I started to sing songs of praise using these words. That night I went to bed, that wonderful feeling of happiness still within me, to wake up still carrying it the next morning. It felt as if I had a piece of heaven implanted in my heart. The experience significantly increased my devotion for Jesus. I was filled with a strong wish to live my life according to his will ever after.[1]

Another of my informants, whom I named Lilly, wrote that

> I was sitting at the kitchen table when the following happened. It felt as if I was one with the whole universe, in other words, one with the whole of God's creation and it felt as if I was in God and the Godship was in me. It was a fantastic feeling and I have never experienced anything like it before or since. I had this strong and certain feeling that God, by way of this experience, called me to start studying theology and to become a priest. There were no words, but I just knew that it was that which happened; that I received my call from God. Shortly afterwards I started studying – the rest is history.[2]

Are these experiences, as my informants remember them, experiences of a Divine Reality or of God, or are they merely a product or function of our nervous system, or a consequence of some brain or psychological disorder? These are some of the questions that shall be considered in this study.

Neuroscientists place different explanations at our disposal concerning what religious experiences are and what causes them. One explanation is

---

1 From one of my informants, my free translation from Swedish, cited with permission.
2 From one of my informants, my free translation from Swedish, cited with permission.

that religious experiences can be explained by studying epileptic seizure experiences, or schizophrenic hallucinations. It is maintained that the human brain itself is able to create experiences of non-material reality and cause experiences of ecstasy or feelings of timelessness. The idea is that the brain, due to certain circumstances, depicts things and events in such an extreme way that persons who undergo such experiences will interpret them as coming from *out of the body*, or as *sacred*. These neuroscientists explain religious experience in terms of consequences of a damaged, mentally deranged or malfunctioning brain. Some neuroscientists maintain, for instance, that Joan of Arc, Saint Paul, Swedenborg, Saint Teresa of Avila, and Saint Birgitta of Sweden, suffered from epilepsy, and that this might have been the cause of their religious experiences and revelations.[3] According to Fraser Watts, the idea that religious experiences are related to epilepsy is the most widely canvassed neurological theory of religious experiences.[4]

The relation between religious experiences and epilepsy seems to be such a pressing issue that novels are written on the topic, for example *Lying Awake* by Mark Salzman.[5] The main character of the novel is a Catholic nun named Sister John of the Cross, who is known for frequently having mystical experiences and revelations. However, her experiences are often accompanied by powerful headaches. The diagnosis is soon made that she suffers from epilepsy. Nevertheless, she can be treated, but the operation will put an end to her religious experiences.

We may find a historical reason why religious experiences are related to epilepsy in early Greek medical literature. The Greek term *epilepsia* means "to lay hold of, seize, attack".[6] Hippocrates, who is said to be the founder of rational medicine, wrote (or was a co-writer of) a medical study of epilepsy entitled *On the Sacred Disease*. Even though the author(s) of *On the Sacred Disease* tried to prove that epilepsy has nothing to do with the *sacred*, epilepsy has more or less been associated with religiosity ever since. For example, the Latin term *morbus sacer* (the holy disease) reveals how epilepsy was commonly understood. One of the reasons is that the layman's idea has long been that only God could throw a person on the ground like this and bring him or her to life again.[7]

Another example of a neuroscientific explanation, more precisely a clinical neuropsychological explanation, is that religious experiences have their roots in existential crises, the most severe being the anticipation of one's own death or the death of a loved-one. The advocates of this explanation

3 Such a list is found in Saver/Rabin (1997) 501–502.
4 Watts (1999) 333.
5 Salzman (2000).
6 Mumenthaler/Mattle (2004) 493.
7 Arts (2001) 29–39, Mumenthaler/Mattle (2004) 493–494.

argue that these crises are the origin of the concept of God because the concept of God, in most religions, is associated with hope, comfort, and eternal life, which is what the person undergoing such a crisis needs. The idea is that when people undergo deep existential crises and are exhausted and anxious, they are more prone to have a religious experience. Furthermore, since they may feel as though they are reborn afterwards and because *being born again* is a feature of salvation in most religions, even people who were not religious at all before the crisis may, after it, join a church or some other religious community.

However, these experiences, according to these neuroscientists, have nothing to do with God or a Divine Reality. What happens is rather that human beings re-experience, as it were, the relation they had with their parents or parent-substitutes during infancy. They think that they experience God but what these persons really do is to mentally answer to their needs in a way that is similar to how they remember their parents answering to their needs.

This psychological principle is called *the principle of association*, which means that our behaviour towards God is equal to how we behaved towards our parents or parent-substitutes during our early childhood. Thus once again, religious experiences are explained in such a way that they have nothing to do with God or Divine Reality.

Nevertheless, there are other psychologists who performed research on the relation between religious experiences and personal crises, but who do not explain them merely in terms of, for example, childhood memories. They do not exclude the possibility that the experiences are actually experiences of Divine Reality. They maintain that religious experiences, whether these are related to a personal crisis or not, are a matter of faith. Often, to them it is enough to know that the experiences were real for those having them.[8]

Furthermore, there are neuroscientists who maintain that religious experiences are correlated with the brain (as all our experiences are) but they do not consider that such correlations exclude the possibility that the experience is caused by God or Divine Reality. Some maintain that experiencing Divine Reality is equal to experiencing everyday reality. They argue that if we consider religious experiences to be nothing but neural activity, we have to consider all our experiences as such.

Can neuroscientists contribute to answering the question of whether religious experiences are *sacred* or *neural*? Can they provide the answer to the questions of whether God exists and somehow, by way of our brain, communicates with us, or whether human beings are, to describe it in the words of the late Francis Crick, "[y]ou, your joys and your sorrow, your memories

8 Geels (2001).

and your ambitions, your sense of personal identity and free will, are in fact no more than the behavior of a vast assembly of nerve cells and their associated molecules. As Lewis Caroll's Alice might have phrased it, 'You're nothing but a pack of neurons'"[9] Consequently, is God merely a product of neural activity? These are issues which will be focused upon.

1.1 The Aim and Key Questions

Sacred or Neural? The potential of Neuroscience to Explain Religious Experience is a critical philosophical investigation of a selection of neuroscientific research on religious experience performed by Michael Persinger and Andrew Newberg and Eugene d'Aquili.

The purpose is (1) to clarify which religious experiences neuroscientists are able to measure by way of today's neuroscientific methods, (2) to analyse and to critically evaluate Persinger's and Newberg and d'Aquili's research that has been performed on religious experiences, (3) to clarify the way in which and to what extent neuroscientists can explain religious experience, and (4) to find a possible way to develop tenable explanatory models for religious experience.

Bertrand Russell once said that philosophy needs the nourishment of empirical science if it wants to avoid stagnation.[10] Although this may be true, it is my conviction that science, and not merely empirical science, needs the nourishment of philosophy as well, and for the same reasons. Both academic disciplines could become too internal, seeing their research only from their own perspective. Michael Lockwood maintains that philosophy "needs to be perceived as having a vital contribution to make to other disciplines, and to human understanding generally."[11] It is my hope that the present research project will reveal the importance of philosophical, theological and neuroscientific disciplines being perceived as having vital contributions to make to one another.

The problem and the core question of this research project is: *In what way and to what extent can neuroscientists explain religious experience?* To be able to answer this question, different neuroscientific explanations and their rationales are identified, what a religious experience is, is elucidated, and what it means to explain something is clarified.

---

9 Crick (1995) 3.

10 Lockwood (1989) 315. Empirical means relying upon or derived from observation or experiment.

11 Lockwood (1989) 315.

1.2 Material and Method

My first idea, which was to give an exhaustive presentation of the neuroscientific research performed on religious experiences, proved to be unreasonable due to the large scope of material that is available. I had to let go of it and instead make a choice. My choice fell on Michael Persinger's work on religious experiences and the research on religious experience done by Andrew Newberg and Eugene d'Aquili for two reasons. Firstly, Persinger, as well as Newberg and d'Aquili, have published their results in neuroscientific and interdisciplinary journals and have written monographs on the topic of neuroscience and theology, which makes it easier to compare and evaluate their studies. Secondly, and more importantly, Persinger and Newberg and d'Aquili offer different and opposing explanations of religious experiences. Where Persinger seems to draw the conclusion that his research undermines the validity of religious experiences, Newberg and d'Aquili seem to draw the opposite conclusion. However, I also consider other neuroscientists' research done on religious experience in order to help me illuminate and critically evaluate Persinger's and Newberg and d'Aquili's studies. To name some of them who play an important role in the discussion: Nina Azari, Jeffrey L. Saver, John Rabin, Peter Fenwick, Stephen Galliano, Carol Rausch Albright, Vilayanur Ramachandran, Michael L. Spezio, Ron Philipchalk, Dieter Mueller, H. Rodney Holmes, Mogens Dalby, and Pehr Granqvist, who actually repeated one of Persinger's experiments on religious experiences.

The method used is a conceptual and argumentative philosophical one as it is typically used in analytical philosophy. The material I utilize consists, on the one hand, of a careful selection of books, anthologies and articles written by neuroscientists concerning religious experiences, but also concerning brain studies in general. On the other hand, it contains a careful selection of work written by theologians, scholars of religious studies, and philosophers who have reacted to the neuroscientific explanations of religious experiences.

As the testimony of Cindy above reveals, I have performed a psychological enquiry on religious experiences. The enquiry was a simple questionnaire with one question being an open one. The aim of the enquiry was to do a brief interrogation on how religious people today interpret their religious experiences, and not to perform a professional psychological investigation. Hence, my query does not fulfil all the criteria that a psychological investigation actually demands in order for it to be scientifically valid. The questionnaire was conducted in different churches in Uppsala, for example, the Swedish Church (protestant),[12] the Baptist Church, the Mission

12 Former Swedish State Church.

Church, the Catholic Church, the Greek Orthodox Church, the Pentecostal Church, but also among Muslim believers. The questionnaire was also sent to established Yogi in order to distribute it to their members. The questionnaire was also circulated among students at the Department of Theology at the Uppsala University. Unfortunately, I did not succeed in distributing it in the Russian Orthodox Church nor in Jewish Synagogues. Three hundred and fifty inquiries were delivered. I do not know how many were copied and circulated internally. However, only 42 questionnaires were returned (23 inquiries were completed by women and 19 by men) and only one interview could be performed. Complete anonymity was promised. The language of the questionnaire was Swedish and the questionnaire was introduced by a letter.

1.3 Interdisciplinarity

That my project would become an interdisciplinary one became obvious when I realised that I wanted to reach neuroscientists, theologians, scholars of religious studies and philosophers. It is meant to contribute to the research field called *science and theology*. Since it can be assumed that not all readers of this research work are philosophers, it may be of help to briefly present philosophy of religion and the research field of science and theology. Furthermore, since not all readers are neuroscientists I shall also give a brief account of neuroscience.

1.3.1 Philosophy and Philosophy of Religion

Talk about philosophy arouses different thoughts. Some see philosophy as a pursuit of wisdom, others as a search for a general understanding of values and reality by chiefly speculative rather than observational means and yet others will regard philosophy as an analysis of the grounds of fundamental beliefs. The task of philosophers of religion is to study problems raised by religion, theology and religious studies by way of philosophical methods. Thus, their field of research concerns different religions and worldviews.

For example, questions of meaningfulness, consequence, soundness, and acceptability, often lead the philosophical analysis of religion. Philosophers of religion study philosophical issues in the religions of the world, in the contributions of philosophical theology and philosophy of religion in western history and in ancient philosophical theology, such as pre-Socratism, Platonism, Aristotelism. An example of such a philosophical issue is found in the theistic concept of God, i.e., in God's omnipotence, omniscience, goodness and omnipresence, etc. and in how such a God could then act in the world. How could this rationally be understood? Another example concerns the justification of religious beliefs, i.e., is it rational to have a religious faith and what good grounds are there for believing such and such? Yet another example concerns the issues of good and evil. For some time now, philosophers of religion have been engaged in the debate of science and religion.[13]

1.3.2 Theology, Religious Studies and Religion

Because neuroscientists refer to theology in their studies of religious phenomena, and because they do not seem to make clear what they mean by theology, I believe it illuminating to present the discipline and its relation to religion. According to Mikael Stenmark, religion is an intellectual enterprise like science only in the form of theology. By his statement, he wants to note that religion and theology are not interchangeable concepts. For Stenmark it is important to consider the differences between theology and religion, especially when comparing the natural sciences and theology as well as the natural sciences and religion. It is not always clear whether the distinction is made and often, unfortunately, theology and religion are blended. According to Stenmark, the crucial difference between theology and religion is that no academic education is needed to practise religion. But when it comes to practising theology, a special training and an academic degree are typically required.[14]

Theology is generally understood as the lore of God or Divine Reality and is a cautious presentation of the context of a certain religious tradition, for example Christianity.

A distinction is often made between non-confessional and confessional theology. Non-confessional theology allows scholars, independent of their own worldview or faith, to study and analyse the context of a certain religious tradition. For example, a scholar who wants to analyse a specific item of the Christian tradition need not be a Christian believer. A requirement for doing confessional theology on the other hand, is that the scholars themselves belong to the religious tradition that is the subject of the study. Consequently, only those belonging to the Christian faith may study Christian issues. Obviously, confessional theology cannot be equated with religious studies because religious studies are the non-confessional study of religious traditions in all their forms and expressions.[15] I shall understand theology as

---

13 For further reading see Quinn/Taliaferro (2002), Guttenplan (2001).

14 Stenmark (2004).

15 Stenmark (1996) 155–156.

non-confessional in the same sense as religious studies. However, I also want to note that some of the people studied below understand theology to be confessional in character.

It is also important to make a distinction between theology and religious studies concerning the scope of their study and research. Inspired by Anders Jeffner's illustrative chart, I suggest that the scope of research for religious studies, similar to psychology, the social and natural sciences and some topics of philosophy is the natural world. The scope of study for at least some forms of theology and some topics of philosophy is both the natural world and the Divine world.[16] Even if it is commonly accepted that the Divine world belongs to the field of study of confessional theology, in my view, non-confessional theology may also have the Divine world as one of its research objects.

Thus, the research field of this kind of theology and sometimes also of philosophy of religion, both concern the natural and the Divine world. Accordingly, theologians and some philosophers of religion study both how people interpret the Divine world and also try to explain the Divine world. The research field of religious studies only concerns the natural world, and studies how people (including philosophers) comprehend the Divine world. Some philosophers of religion, then, affiliate with religious studies.

1.3.3 Science and Theology

As the name of the discipline Science and Theology reveals, academics working within this field are interested in both science and religion in one way or another. They may belong to the academic field of the natural sciences, the social sciences, the human sciences, theology or religious studies, but what they have in common is that they are interested in the relation between science and religion. It was during the 1960s that

> major developments in the philosophy of science and the philosophy of religion, new theories and discoveries in the natural sciences, as well as complex shifts in the theological landscape, made possible constructive interaction between often separate or even hostile intellectual communities.[17]

One could perhaps say that it was Ian Barbour with *Issues in Science and Religion* who built the first bridge between the two fields.[18] He accounted for the mutual aspects he found in the methodology, linguistics and epistemology of the two disciplines of science and theology, namely "[b]oth

---

16 Jeffner (1975) 14.
17 Russel/Wegter-McNelly (2003) 746.
18 Barbour (1996).

made cognitive claims about the world expressed through metaphors and models, and both employed a hypothetico-deductive method within a revisionist, contextualised, and historicist framework".[19] Thus Barbour maintained that what both disciplines have in common is that they need to work with hypotheses, metaphors or other figurative language, explanatory models, etc. to construct theories about our world.[20]

1.3.4 Neuroscience

Neuroscience is, roughly, a branch of the life sciences that deals with the nervous system and its functions. Neuroscience has its roots in early Greek medicine, and thus has a long history. However, for the present study, suffice it to know that it was during the 19th century that neuroscientists discovered that nerve energy is actually electrical energy and not, as was believed until then, something *spiritual*. Neuroscience also discovered that the neuron is the basic component of the brain, a biological cell specially adapted with axons and dendrites to transmit neural information to other cells. Furthermore, they discovered that this neural system is similar in all creatures, because whether it is a primitive creature like a flat worm, or a human being, the neural system works along the same basic principles of chemical stimulation and electrical conduction. The difference between the species lies in the complexity of the neural system rather than in its function.[21]

Also, during the 19th century, speech, language and motor modes received their specific locations within specific areas of the brain: Among others, the somatosensory area was located in the parietal lobe, the visual projection area was located in the occipital lobe and the auditory projection area was located in the temporal lobe.

In the 20th century, as the possibilities for experimental studies increased, neuroscientists maintained that kindling (electrical stimulation) of the brain produces immediate phenomenal experiences like epileptic seizure experiences. Furthermore, the novel possibility of separating the hemispheres surgically (cutting the corpus collosum) along with psychological studies, among which was the well known *blind-sight* study, led to the conclusion that phenomenal awareness of events is utterly dependent upon the viability of certain neural connections. Hence, neuroscientific research seemed to conclude more and more that mental activity is connected to neural activity.

---

19 Russel/Wegter-McNelly (2003) 746.
20 By hypothesis I shall in this study mean a proposed explanation based on limited evidence and hence used as a starting point for further investigation.
21 Newberg et al. (2001) 15/177.

Contemporary neuroscience is a multidisciplinary research field containing disciplines such as neurobiology, neuropsychology, neuroanatomy, neurophysiology, and neuroradiology. Warren S. Brown groups neuroscience into the following academic disciplines: Cellular and molecular neurophysiology; developmental neuroscience; sensory and motor neuroscience; regulatory neuroscience; behavioural and cognitive neuroscience; clinical neuroscience.[22] It is to be noted that neuroradiology belongs to the discipline of clinical neuroscience. We shall see more about it in the proceeding chapters.[23]

1.3.5 Neurotheology

How are neuroscience and theology related? During the last 20 years, we have witnessed an explosion of inquisitiveness for the psychological and the spiritual. Moreover, Newberg writes that

> The neurosciences clearly have significant potential in helping with the understanding of religious experiences and religious phenomena. An array of imaging technologies as well as the ability to measure various physiological and neurophysiological parameters will likely have a significant impact in terms of how we come to understand the importance of religion and religious experiences in human life.[24]

This has led to a new research field called *neurotheology*. According to Palmyre Oomen, there are two main streams of neurotheology.[25] She maintains that the first stream, introduced during the 1970s until the 1980s, involved d'Aquili and Laughlin (among others) and related neuroscience to religious behaviour such as rituals and religious experiences by way of looking for their neuropsychological determinants. The second stream, according to her, had its start in the 1980s. The focus lay more and more on the search for correlations between religious experiences and specific brain structures and functions. Here we find, according to her, Michael Persinger and Andrew Newberg. Newberg and d'Aquili worked together until d'Aquili passed away, shortly before the publication of *Why God Won't Go Away*, in 2001. In the second period of neurotheology, the method changed from being neuropsychological to neuroclinical, especially, neuroradiological.[26]

In my view however, the second stream of neurotheology can be understood in different ways. Firstly, it can be understood as neuroscientists

22 Brown (2003c) 611.
23 Brown (2003c) 610–617.
24 Newberg (2002) 145.
25 Oomen (2003) 617–618.
26 We shall become acquainted with a part of how this process of change was established in chapter four, when the work of Newberg and d'Aquili is presented and analysed.

engaging in neuroscientific problems that have to do with theology and religion. Neuroscientists want to know whether God or other religious phenomena and religious behaviour can be explained neuroscientifically. They investigate religious issues (for example religious experiences) in order to explain them in neuroscientific terms. When they succeed, religious issues are no longer a problem for science because they *show* that it is the brain that generates religious experiences and nothing else. It is this type of neurotheology that involves, among others, Persinger.

A second way of understanding neurotheology is, to put it in Newberg's own words: "Bringing 'Neuro' and 'Theology' Together Again."[27] The idea is that neuroscientists investigate how religious phenomena and behaviour can be understood from a neuroscientific point of view (which denotes the *Neural* part of the term), taking the theological and religious aspects into consideration (which denotes the *theological* part of the term).[28] The focus does not lie in giving religious phenomena neuroscientific explanations but rather on understanding the correlations between religious phenomena and what neuroscientific studies have been able to show concerning these religious phenomena.

There might be another possible understanding of neurotheology. From this perspective, neurotheology is the search for understanding the correlation between religious phenomena and what a cooperating research team from different disciplines (but always including the neurosciences and theology), have been able to account for concerning these religious phenomena, in order to find a tenable explanation. Such a research team could, it is suggested, consist of theologians and scholars from religious studies; philosophers of religion; neuroscientists (not least neuroradiologists); in addition, ethicists, ethnologists and anthropologists.

1.4 Preview of the Chapters

Generally, in chapter two the aim is to construct a philosophical toolbox in order to be able to analyse and evaluate the work of Persinger and Newberg and d'Aquili and to be able to answer the different questions I posed. It is chapters three, four, five, and six that include the actual philosophical analyses and evaluations. In chapter seven then, the answer to the main question is provided.

In chapter two, which, as mentioned above, figures as a philosophical toolbox, the differences and similarities between non-religious and religious

27 Newberg (2002) 145.

28 Newberg and d'Aquili were not attentive to the difference between theology and religion.

experiences will be specified, as well as their relation to human experiences. It will also be investigated what it means to explain something. Are religious and non-religious experiences different interpretations of human experiences or are these experiences significantly distinct from one another? What exactly is a religious experience? Since not all religious traditions define God or Divine Reality in a similar way, the question is also raised how claims about God or Divine Reality should be understood. Furthermore, a distinction will be made between psychical experiences and religious ones, because not all neuroscientists pay attention to the differences between the two. To explain is always to explain something and the purpose is always to clarify something, to clarify a specific item of something, to make something easier to grasp. However, there are many ways of making things intelligible: By way of their causes; by way of how-explanations; by way of how things are or occur; by way of why-explanations; by way of categorizing things and events or by identifying them. Generally we blend possible ways of explaining and we do not really reflect on the plausibility of our explanations. When it comes to academic disciplines however, we have to distinguish between real and spurious understanding, because of the expectation that an objective explanation, i.e., which is in accordance with the criteria established by science, will be presented. This has led the natural sciences to apply different models. However, theology and religious studies also apply models to explain things and events. Obviously, the differences and similarities in methods of explaining between the natural sciences and theology and religious studies will be clarified. Finally, I will present what, according to me, it takes to explain the experiences of others.

In *chapter three*, I shall interpret and clarify Michael Persinger's neuroscientific studies of religious experience. Persinger's research is meant to give him answers to the following questions: Are religious experiences what they are claimed to be or are they rather the result of different derailments of the brain? If the latter is true, can we know which neurological processes generate religious experiences? The chapter will start with a presentation of the neuroscientist. The analysis will concern: Persinger 1987, 1993b, 1993c 1997 and 2003. My investigation will be limited to his studies of religious experience and, therefore, those studies in which he investigates psychical experience, UFO's, or experiences of poltergeists, will not be considered.[29] I believe that the analysis will provide us with a good idea of which religious experiences Persinger investigates; insight into how his research was performed, and it will allow me to further evaluate his

29 However, since Persinger considers all these experiences to be religious ones, it is difficult to make a clear distinction.

work. It shall become clear whether and to what extent I think we should agree with Persinger.

Chapter four will be devoted to the evaluation of the work of Persinger that is analysed in the previous chapter. Since I believe that it is important to investigate how his research on religious experience is performed from a scientific point of view, I shall embark on the chapter by evaluating the scientific validity of Persinger's studies. In other words, it will be explored whether and to which degree these studies measure what they purport to measure or what Persinger believes his research measures. The chapter will be continued by evaluating other aspects of Persinger's studies. For instance, his idea that religious experiences are somehow related to epilepsy will be compared with other research done on the relation between religious experience and epilepsy. Also, his studies of the implications of religious experiences on the behaviour of men and women will be compared with other studies in order to see whether these studies could support Persinger's. For example, has Persinger support for his assertion that right-handed men are potentially dangerous when they have a religious experience? In one of his studies, Persinger searches for a way to isolate the neural processes that generate religious experiences, because he believes that the results of such isolation may be important for clinical use. Could it be as Persinger maintains, that if he could understand the neural basis of religious experiences and hence these could be generated experimentally, they could be a powerful tool in the treatment of depression, anxiety, and dissociation? Also, are religious experiences a consequence of existential crises as Persinger maintains in yet another study? The chapter will end with a consideration of which ways of religious experiencing and which types of religious experience Persinger was able to study.

In *chapter five* the task is to interpret and clarify Andrew Newberg and Eugene d'Aquili's neuroscientific studies of religious experience. Their aim is to find which correlations can be established between religious experiences and neural activity. Their goal, it seems, is not only to develop a theory but even to explore all of the implications for theology, philosophy, and the sciences. The chapter also begins with a presentation of the neuroscientists. The studies which I considered to be the most important for the purpose of my project and which will be presented and philosophically analysed are, Newberg et al. 2001a and 2001b, d'Aquili and Newberg 1993a, 1993b, 1999 and 2000. The focus will be on their neuroscientific models for religious experiences obtained by way of different meditation techniques that show which brain functions and structures are expected to correlate with the different religious experiences along the meditation process. These models might be considered to have inspired Newberg and d'Aquili to undertake their experiment. Then, the conclusions that Newberg and d'Aquili have drawn from their research will be analysed. These are, among others, that there is an invariant element across the religions, that all experiences are equal and that God is real. Similar to the analysis of Persinger's work, I believe that the analysis will provide us with a good idea of which religious experiences Newberg and d'Aquili investigate; insight into how their research was performed, and it will allow me to evaluate whether their results and the conclusions they draw are plausible. It shall also become clear whether and to what extent I think we should agree with them.

In *chapter six*, I shall evaluate the work of Newberg and d'Aquili. The evaluation, once more, starts by investigating how the neuroimaging experiment on religious experiences is performed from a scientific point of view. Furthermore, the results of the neuroimaging experiment will be compared with Newberg and d'Aquili's models for religious experiences obtained by way of meditation, not least because this has some impact on the content of chapter eight. One purpose is to investigate what their research can and cannot tell us concerning religious experience and to what extent the experiment confirms the models. Moreover, we shall see whether Newberg and d'Aquili were able to find the invariant element across the religions which they were looking for. Moreover, is it the case that all experiences are equal neurologically? Also Newberg and d'Aquili's idea that God or Divine Reality is real shall undergo an evaluation. An implication of one of the results of the neuroimaging experiment was that meditators and non-meditators had different thalamic values at baseline. What does this result mean? Finally, also here, the ways of religious experiencing and the types of religious experience that are studied by way of neuroscientific methods will be discussed.

Chapter seven will provide us with the answer to the core question *in what way and to what extent can neuroscientists explain religious experience*? Hence, relevant reductionist views will be presented. Since reducing religious experiences can be done in an erroneous or in an adequate way, these variants will be discussed in order to see in which ways Persinger and Newberg and d'Aquili apply reductionism. Moreover, it will be investigated whether and how different explanations of religious experience may be combined. The purpose is to see what kinds of explanatory models Persinger and Newberg and d'Aquili use and to evaluate if and how these models could be further developed.

The last chapter, *chapter eight* will contain some final conclusions and suggestions to neuroscientists who are interested in empirical studies of religious experience.

2 Experiencing and Explaining

I am certainly not the first scholar to ponder upon what it is to experience and what it is to have a religious experience. Still it is my hope that my thinking will add some novelty or at least another perspective to the topic. Anyway, I have to tackle the problem of experiencing if I want to put forward an answer to the question *in what way and to what extent can neuroscientists explain religious experience*. Hence, I shall investigate if and in what sense non-religious experiences and religious experiences differ from one another and how their relationship with human experiences should be understood, as well as, investigate different ways in which one can experience something and have a religious experience. It is important to keep in mind that the intention is not to present an exhaustive explanation of what it is to have an experience or how religious experiences may be understood. I shall not become engrossed in the biology of the brain concerning what is required for a human being to be able to have an experience, nor shall I enter deeper into a discussion concerning whether a human being has to be conscious or not to have an experience. Instead, I leave these interesting issues aside and will concentrate on what it means for a healthy conscious human being to have an experience.[1] Nor shall I enter into the debate concerning whether or not God really is *out there*. Rather, my intention is to suggest understandings of these issues that are fruitful for the task of answering to the core question, i.e., to construct a toolbox.

2.1 Human Experiences

What is a human experience? What does it mean to have an experience and in which way(s) does a human being have experiences? These are some of the questions to be answered.

By human experiences I intend all experiences a human being is able to have. Since a human being is related to the world, I shall designate a human experience as an experience of things and events whose existence ultimately is constituted of properties and relations, actions and interactions of

---

1 A conscious human being, in my understanding, is a person who is aware of his or her being alive in a world, everything having to do with this being involved.

whatever the physical and psychological worlds treat of.[2] Add to it that an experience should be understood as an expression of a relation between the experiencer and the experienced.

What does it mean to have a human experience? At this moment, I am aware of the sun shining through the window, the birds singing outside that same window of the office where my desk is situated and where I am writing on my computer. I am aware of the buzzing coming from four other computers around me. I am aware of my irritation concerning that matter and of my trying not to pay attention to it. I am aware that I *experienced* all this between 11.05 am and 11:13 am. In other words, I had an experience of the sun shining through my window, of the birds singing, etc. What does this mean?

According to Keith Yandell, a (human) experience, let us call it *x*, is *an aspect of the subject of experience*, let us call it *y*.[3] Yandell maintains that "if an item *A* is so related to person *S* that *S does not exist* by itself or together with some set of truths, entails *A does not exist*, then *A* is an *aspect* with respect to *S*".[4] He argues that every human being is an aspect with respect to her- or himself but not to another person. My hopes and dreams, beliefs and attitudes are aspects with respect to myself, but objects with respect to you. It is important to note that the subject of experience is not the person who experiences (the experiencer) but is that of which an aspect relative to the experiencer is presented to him or her. Yandell continues that "[h]aving an experience is an aspect of the experiences' subject. Experiences all have subjects, though not all have objects. "That *having an experience* is an aspect [of the experiences' subject] relative to the one who has it [...] does not entail that experiences cannot have objects."[5]

Yandell thus argues that all human experiences are aspects of a subject but do not necessarily have an *object*, let us call it *y*'. For example, the object of my experience mentioned above *x*, has an aspect of the subject of my experience *y*, namely, the sunshine through my window and even an object, namely, the sunbeams *y*'. Hence, an experience *x* consists of *y* and sometimes, but not necessarily, even of *y*'. It is important to note that it is not the case that an experience is either an aspect of a subject of experience or an object.[6] An illustration of a human experience that does not have any object would be experiencing thirst, because there is no object in that experience. The experience of being thirsty is an aspect of the subject of experience, namely, neural signals from the central nervous system. Experiencing drinking water on the other hand has an object, namely, water.

---

2 Lockwood (1989).
3 Yandell (1994) 42.
4 Yandell (1994) 38.
5 Yandell (1994) 42.
6 Yandell sometimes does not articulate this himself.

Yandell also maintains that "*[a]n* experience is a conscious state in which a subject [a person] (phenomenologically) seems to be aware of an aspect or object".[7] The term *seems* applied by Yandell is, as it will become clear, important for our discussion. By *seems* he means "how things appear, whether or not they are that way or not".[8] This remark is important, because it means that for an experience of something (an aspect of a subject of experience with or without an object) to be recognised as an experience of that something, it does not necessarily imply that this something really is as it is experienced. A person who is afraid of snakes and sees the tail of a cat wagging from behind some stones, thinking it is a snake, will experience her or his fear of snakes. The experience of fear will be recognised as such until the person finds out that the snake she or he seemed to have experienced was actually the tail of a cat. In other words, the experiencer will know whether her or his experience of fear is legitimate when she or he receives evidence that justifies her or his experience of fear.

However, even if a person has an experience that is directly sustained by the evidence needed for some claim, the person might not know that. For example, the person might actually have seen a snake but was too far away to really know this. Moreover, the person might know that there is direct evidence for her or his claim but might not see it as evidence for her or his claim, because the person might take her or his experience as not being veridical. This also makes sense, because the person might, for example, be so intensely afraid of snakes that she or he simply refuses to accept her or his experience of one. Yandell distinguishes between "*de facto evidence*" and "*effective evidence*" and maintains that an experience *x* is de facto evidence for a claim *c* if the experience *x* meets the relevant conditions (those conditions making the experience legitimated).[9] An experience *x* is effective evidence if it is both de facto evidence and accepted as such by the experiencer. An interesting effect of Yandell's distinction can be illustrated by the following thought experiment. Assume that Preben actually saw a snake, i.e., that his experience of fear of snakes met the relevant condition that there was a snake behind the stones. This means that his experience is de facto evidence for there being a snake. However, Preben does not accept this and maintains that he has a phobia of snakes, has hallucinations about them and his experience was just another hallucination. Hence, his experience does not count as effective evidence. However, assume that Julia, Preben's wife, knows that Preben actually saw a snake, because Preben always represses his experiences of snakes and because it is not the case

7 Yandell (1994) 42.
8 Yandell (1994) 16.
9 Yandell (1994) 44.

that Preben suffers from hallucinations. To Julia, then, since she sees Preben's experience as evidence for there being a snake behind the stones, Preben's experience would be effective evidence. Thus, a de facto experience may be accepted as such and count as effective evidence for a person who did not undergo the experience him- or herself while the person having the de facto experience does not recognise it as such. This distinction made by Yandell will be shown to be useful for our discussion. According to Yandell, these conditions are sufficient for having an experience.

I believe Yandell's criteria for what it means to have an experience to be fruitful for my aim of constructing a defensible understanding of religious experience and I shall therefore apply and further develop some of his thinking. For example, I would like to add some conditions which I believe are important. One such condition is that the aspect of the subject and the object of the experience is always presented to the experiencer in a certain way. Alston, for instance, writes that "to perceive a house is for a house to be directly presented to one's experience".[10] The world is presented to the experiencer, thus the experiencer is passive concerning the subject of experience.

2.1.1 Ways of Experiencing

However, there are different ways in which a person could have experiences.[11]

According to Peterson we have experiences in which we actually *participate* (like playing chess) and those in which we take part as *witnesses* (like watching somebody playing chess).[12] Yandell distinguishes between *perceptive* experiences in which we are aware of an object relative to us, and *introspective* experiences in which we are aware of an aspect (of the subject of experience) relative to us.[13] It is important to keep in mind that introspection, as Yandell defines it, should not be understood as the act of *looking into* one's own thoughts and feelings. To Yandell, an "*[i]ntrospective* experience, in contrast to perceptual, concerns aspects rather than objects".[14] Finally, Alston makes a division between having *sensory and non-sensory*

---

10 Alston (1991) 28.

11 I shall however, not give a presentation of the neurological basis for how we experience. The reason is simply that it is not possible, because experiencing is a very complex neurological issue. The neural structures involved are far too many, those involved in perception, in cognition, in movement, etc. For my purpose, on the other hand, it is interesting to present different ways of experience.

12 Peterson et al. (2003) 16.

13 Yandell (1994) 42–43.

14 Yandell (1994) 43/38/280.

experiences. His idea is that in sensory experiences something is literally presented to us in contrast to non-sensory experiences where there is a specifiable mode of consciousness through which something is perceived.[15] Alston's non-sensory concept can be understood as a *looking into* one's own thoughts and feelings. Since God is pure spirit, God will probably present Godself in a non-sensory experience rather than in a sensory one, he says. It is, however, not really clear what he means by senses. I in my opinion, it is unlikely that he includes intuition among the senses, because intuition, rather, is a cognitive feeling and belongs as such to the non-sensory way of experiencing. Even if it is possible to construct eight combinations of the views of Peterson, Yandell and Alston, only six are adequate:

1. The *participating perceptive sensory way* of experiencing.
2. The *participating introspective sensory way* of experiencing.
3. The *participating introspective non-sensory way* of experiencing.
4. The *witnessing perceptive sensory way* of experiencing.
5. The *witnessing introspective sensory way* of experiencing.
6. The *witnessing introspective non-sensory way* of experiencing.

Those that are not applicable are the participating perceptive non-sensory way of experiencing and the witnessing perceptive non-sensory way of experiencing. This is because a perceptive experience implies that there is an object while there are no objects in a non-sensory one. We thus have a contradiction.

What could these ways of experiencing look like? To begin with, playing football, tennis, dancing and playing chess are good examples of a *participating perceptive sensory way* of experiencing for the following reasons: (1) The experience of playing implies with necessity that one participates (in contrast to merely watching others playing). (2) The experience is perceptive, due to the objects that necessarily are involved, a football, tennis balls, one's own and others' bodies, fields, chess pieces, etc. (3) The experience is clearly sensory (seeing where you put your feet when dancing).[16]

Solving a mathematical problem on the blackboard could be an example of a *participating introspective sensory way* of experiencing. (1) Solving a mathematical problem surely needs your participation: you cannot merely witness you are solving the problem. (2) It is introspective, because mathematics is not something physical: mathematics does not have objects. (3)

15 Alston (1991) 19–23.

16 Blind people are of course able to do all that as well, relying on senses other than the visual senses.

However, using the blackboard instead of solving the problem in your mind involves some of your five senses.

Sitting in a rocking chair with your eyes closed and mentally trying to solve that mathematical problem would turn your experience into a *participating introspective non-sensory way* of experiencing. Since you solve it in your mind (and not on the blackboard) none of the five senses are involved: everything else remains the same.

Watching my son playing football would be an example of a *witnessing perceptive sensory way* of experiencing. (1) You might, for example, pay attention to how he plays, how fast or slow he runs, how often he kicks the ball, etc. Hence, you witness, because you are not playing yourself. (2) The experience is a perceptive one, because the football game which my son is involved in, consist of objects related to you as you sit there and watch. (3) Since you have to use several senses, your experience is a sensory way of experiencing.

Experiencing the beauty of an equation on the blackboard is an excellent example of a *witnessing introspective sensory way* of experiencing. (1) You witness the formula on the blackboard. This time you do not write it down yourself. (2) The experience is introspective, because as mentioned above, mathematics does not have objects and neither has beauty (remember, you do not experience the blackboard). (3) It is sensory, because you see it written on the blackboard and not merely in your thoughts.

Finally, we find a distinguished example of a *witnessing introspective non-sensory way* of experiencing in Marcel Proust's *A la recherche du temps perdu*, (Remembrance of Things Past), I quote:

> [M]y mother, seeing that I was cold, offered me some tea, [and] sent for one of those squat, plump little cakes called 'petites madeleines' [...] And soon, mechanically, [...] I raised to my lips a spoonful of the tea in which I had soaked a morsel of the cake. No sooner had the warm liquid mixed with the crumbs touched my palate than a shudder ran through me and I stopped, [...] And suddenly the memory revealed itself. The taste was that of the little piece of madeleine which on Sunday mornings at Combray (because on those mornings I did not go out before mass), when I went to say good morning to her in her bedroom, my aunt Léonie used to give me, dipping it first in her own cup of tea or tisane. The sight of the little madeleine had recalled nothing to my mind before I tasted it; [...] But when from a long-distant past nothing subsists, after the people are dead, after the things are broken and scattered, taste and smell alone, [...] remain poised a long time, [...] and bear unflinchingly, in the tiny and almost impalpable drop of their essence, the vast structure of recollection.[17]

This frequently quoted experience triggered by eating the petites madeleines is (1) a witnessing way of experiencing, because Proust is not partici-

17 Proust (1913) 48.

pating in the sensation which the eating of the petites madeleines arouses, but witnesses his past, (2) an introspective way of experiencing, i.e., there is no object involved, because he does not experience the madeleine cookie, nor eat it, but experiences the memory of his aunt Léonie dipping the madeleine cookie in her tea cup and (3) a non-sensory way of experiencing: Even though it was the smell and taste of the madeleine cookie that triggered the experience, these are not the experience itself. Observe that this witnessing introspective non-sensory experience has evolved from a participating perceptive sensory experience, namely from eating a petites madeleine cookie (the cookie's smell and taste are, as it were, the *channels* that leads him to the experience of his past).[18]

2.2 Religious and Non-Religious Experiences

The aspect of the subject and object of the experience is hence always presented to the experiencer in a certain way. According to Alston, religious experiences are those experiences in which it seems that God appears (presents Godself) to a person in a certain way. Thus, it seems that Alston places the burden on something (taken by the experiencer to be God) presenting itself (Godself) to a person's experience. Doing so has the advantage of being able to compare experiences of God with other experiences taken by the person to be an experience of something specific. Alston wants to dissent from the idea that all experiences of God and God-agents are nothing but subjective experiences.[19] As already mentioned above, he makes a distinction between sensory and non-sensory experiences. Concerning religious experiences, a non-sensory experience means that God presents Godself to the person as "loving, powerful or good." Furthermore, according to him, subjective features of the experience are involved, "there is an 'ineffable joy and exultation', it involves a 'calm, a repose, and an interior sweetness', and so on".[20] A sensory religious experience would be experiencing Jesus in human form: "[A]ccording to the Christian doctrine, Jesus of Nazareth is both man and God; so that to see him, even in human form, is to see God".[21] Alston's description of religious experiences is propitious for my purpose of finding a plausible definition of religious experience. However, one problem with Alston's definition of religious experience, in the sense that *God appears* to a person, is that the idea that God

18 This remark will be helpful when we shall consider ways of experiencing during meditation sessions.

19 Alston (1991) 14–20. Alston supports critical realism.

20 Alston (1991) 20–23.

21 Alston (1991) 20.

appears may lead to misunderstandings. It may be comprehended as only covering theistic religious traditions, because God could be understood as a personal God distinct from the world. Not all religious traditions define God in a similar way. For example, there are those who define God in terms of the absolute ground of everything that is (and ever may be) and those who comprehend God as the creator of everything that is. There are those who define God as being equal to the world, those who consider God to be totally distinct from the world, and those who maintain that God is both distinct from and equal to the world.[22] Understanding God as a personal God distinct from the world would make the definition of religious experience too narrow for my purpose.

To avoid a narrow understanding which the concept of God might entail, I suggest that we understand *God* as *Ultimate Reality* in the sense that John Hick intends, namely, as "absolutely real, [and] which corresponds sufficiently to the Sanskrit *sat* and the Arabic *al Hagg*".[23] Hick, defending pluralism, searches for a description of God or Ultimate Reality that all religions could agree upon. However, inspired by Paul Tillich's understanding of God, as "the Ground of Being",[24] and Niels Henrik Gregersen's idea that God is "the power in everything that exists",[25] I propose further that Ultimate Reality is understood as *the absolute ground* of everything that is (and ever will be).

Ultimate Reality should be understood differently from ordinary reality, but both ordinary reality and Ultimate Reality as not dependent on observation. Reality (as used in Ultimate Reality) is taken to be the absolute ground of everything that is, whether or not it is observable, accessible or understandable by science, philosophy or any other system of analysis. Reality (as used in ordinary reality) on the other hand, is restricted and is (in part) accessible to scientific and philosophical observation, investigation and theoretical interpretation.[26]

---

22 There may be additional understandings of God and different understandings might even be combined. It is not my purpose to have an exhaustive presentation on how God could be understood.

23 Hick (1996) 47.

24 Thiselton (2002) 307.

25 From our conversations during my stay at the department of Systematic Theology at Århus University, December 2003.

26 Even if God and Ultimate Reality could be considered to be synonyms, both terms will be used together to avoid misunderstanding. By Ultimate Reality I do not mean *Brahman* who is the *Ultimate Reality* by contrast to *Atman* which means *Self* in the Hindu religious tradition. It should be noted that theologians and philosophers of religions to whom I refer below may not understand Ultimate Reality as I have suggested here. However, my purpose is not to present different understandings of Ultimate Reality, but rather to be able to clarify the way God or Ultimate Reality is comprehended by the neuroscientists of my investigation and to see whether they are consistent in their view.

Religious experiences, then, are those experiences in which it seems to the experiencer that *God or Ultimate Reality* appears to her or him in a certain way.

Let us now consider this understanding of religious experiences along with Yandell's statement that a religious experience is "an experience doctrinally and soteriologically central to a religious tradition".[27] However, according to Theologian Sven-Erik Brodd, the relation between doctrines and soteriology is not always made clear and is therefore problematic. Sometimes soteriology is included in doctrines and sometimes it is not. Nevertheless, even if the soteriology of religious experiences sometimes (but not always) is included in the doctrines of the religions, I believe it to be more illustrative to have both mentioned. It is not clear why Yandell chose to define religious experience as such. One reason behind his definition of religious experience might be that he thinks that "there are religions and religious experiences that are implausible [and] even evil".[28] Thus, the reason might be that he wants to be able to exclude evil and implausible religions from his understanding of religious experience. If we accept Yandell's definition of religious experiences it could give us a tool, or part of a tool, that besides broadening our understanding of religious experiences, also gives us the possibility of comparing different religious experiences and of seeing whether the experiences studied by the (neuro)scientists could actually be classified as religious ones. Yandell does not clarify what he means by *central* to a religious tradition, which is problematic, because it could be understood as *essential, vital, fundamental or cardinal.* I propose that for our purpose, we understand *central* as *essential*, because even if this means that the doctrines and soteriology belong to the very nature of a religious tradition and thus cannot be removed without drastically changing the religious tradition itself or its character, it does not exclude the possibility of continuity (see further below). To understand central as vital, fundamental or cardinal, in my view, would be too strong and, if not exclude, very much complicate the possibility of continuity.

However, there are problems with Yandell's view as well. One is that the definition does not seem to leave room for innovative religions. Another problem is that it seems to exclude the value of the interpretation of the person who had the religious experience of what he or she has experienced. There is a third problem related to this. How, for example, could a religious experience that cannot be identified by the doctrines and soteriology of the religions be accepted as a religious experience? Thus, Yandell's idea of a religious experience is in need of some development.

---

27 Yandell (1994) 15.
28 Yandell (1994) 15.

The first problem might be solved if we consider what Yandell means by religion. He writes that

> a 'religion' or a 'religious tradition' is 'a conceptual system that provides an interpretation of the world and the place of human beings in it, that builds on that interpretation an account of how life should be lived in that world, and that expresses this interpretation and life-style in a set of rituals, institutions, and practices'.[29]

Accordingly, we can establish that all religions and religious traditions, including indigenous religions, but not evil and implausible religions, are embraced, because Yandell comprehends religions and religious traditions as conceptual (metaphysical) systems. If we add to his definition of religion and religious traditions the arguement of continuity, i.e., there is a possible continuity between existing and novel religions and religious traditions, I believe the first problem to be solved. Understood as such, new religions and religious traditions satisfying Yandell's criteria for being a religion or religious tradition mentioned above, can be included in his definition of religious experiences, because they then belong to, as it were, the family of religious traditions.

The second problem, that Yandell's definition should exclude the value of the interpreted religious experiences could, in my view, be solved if we add to his definition that for an experience to be a religious one, the experiencer has to recognise it as such. However, what do we mean by religious? Different criteria could account for the religiosity of the experience, for example, that the experience is "an experience of something that is religious",[30] according to the definition, and/or that it is "an experience of something in a religious way",[31] according to the definition, and/or that the experiencer is a religious person. We have seen that an experience should be regarded as an expression of a relation between the experiencer and the experienced. From this it follows that for an experience to be religious, this *relation* has to be *religious*. It is thus not God or Ultimate Reality that is religious, it is not the experiencer who is religious: the religious feature emanates in the interface between the two. In my view then, this implies that an experience is not a religious one unless the experiencer *recognises* his or her experience as such. This is an important addition. For instance, even if a person describes her or his experience in terms of a union with an intensive light taking the form of a cross (which would be in accordance with the doctrines and soteriology of the Christian tradition), it is still not a religious experience as long as the person does not accept it as such. She or he may maintain that what she or he experienced was simply an after image

29 Yandell (1994) 15.
30 Bråkenhielm (1985) 28.
31 Bråkenhielm (1985) 28.

due to the strong reflections of the sun on the water. On the other hand, the fact that a person takes her or his experience as an experience of God does not guarantee that she or he really had a religious experience.

Both Alston and Yandell share my view that a person has to accept her or his experience as a religious one for it to be one. We already mentioned Yandell's idea when presenting human experiences, namely, his distinction between de facto evidence and effective evidence. Accordingly, a religious experience *x* is de facto evidence for a claim *c* if the experience *x* meets the relevant conditions (in the doctrines and soteriology of the religions). A religious experience *x* is effective evidence if it is both de facto evidence and accepted as such by the experiencer. Hence, the experiencer might not recognise the evidence for his or her religious experience (the cross on the water). Furthermore, a de facto religious experience may be accepted as such and thus count as effective evidence for a person who did not undergo the experience him- or herself while the person having the de facto experience does not recognise it as such.

Alston's idea is that

> [t]here can be genuine awareness of God that the subject [the experiencer] does not take as such. First, no such taking may occur, because the subject is not attending to the matter, though she would so construe the experience if the question arose. A second possibility is that one may actually be experiencing God without even being disposed to identify the object of the experience as God if the question arose, just as one can see a cyclotron without realizing that what one sees is a cyclotron. I may be aware of God's sustaining me in being, while I suppose that I am merely feeling particularly fit and chipper at the moment, or I may be 'hearing' God speak to me (not with audible words), while I take this to be just thoughts floating through my mind.[32]

It is the second possibility that is of concern to us. For instance, how many people know what a cyclotron looks like and thus how many would recognise it as such when seeing one? Thus, the person might not know that it was God who presented Godself to him or her. But it may also be the case that the person does not want to recognise that it was God presenting Godself to him or her.

The third problem concerns the possibility that a person may have a religious experience that cannot be placed within the doctrines and soteriology of the religions. If we understand God as the absolute ground of everything that is and ever will be, then God could present Godself to a person in a way that nobody has ever experienced and no religion has yet thought about.

A final adjustment I want to make to Yandell's definition is that I shall not refer to *a* religious tradition as Yandell does, because in my view, living

32 Alston (1991) 11–12.

in multicultural environments as we do today means that the borderlines between different religious traditions are not as articulated as they once were. For example, today, the Buddhist religious Zen doctrine is accepted and applied along with Christian doctrines by some churches and monasteries of the Christian tradition. Furthermore, I suggest that we should take into account the seemingly growing endeavour towards ecumenism.

Consequently, I believe the problems to be solved when we add the arguement of continuity and the criterion that the experiencer has to recognise his or her experience as a religious one (giving the experiencer authority) to the definition that a religious experience is an experience doctrinally and soteriologically essential to the religious traditions.

We also need to investigate how we can understand perceiving something so that perceiving God or Ultimate Reality can be included. Hence, we need a theory that makes perceiving God or Ultimate Reality possible, if God or Ultimate Reality exists. Such a theory is Alston's *Theory of Appearance*.[33] It states that "the notion of X's appearing to S as so-and-so is fundamental and unanalysable. [...] For S to perceive X is simply for X to appear to S so-and-so".[34] Thus, my perceiving the flowers in the garden as red roses is fundamental and unanalysable because of the flowers appearing to me as red roses. We should understand unanalysable as "not analyzable in terms of conceptualizations, beliefs, takings or anything else of the sort".[35]

However, what if these flowers are not roses at all? I might have been too distant from the garden in question to be able to distinguish roses from other flowers. I might have been too hasty in my judgement since I favour roses and *see* them everywhere. However, Alston maintains that the Theory of Appearance allows for the possibility that "what an external object presents itself as may diverge from what that object actually is".[36] The perception of God or Ultimate Reality as so-and-so may diverge from how God actually is, for example, how Moses perceived God as a burning bush.[37]

Nevertheless, an important remark is that the perception of God as so-and-so might also be a perception of something completely different. The person might have made a wrong interpretation. The Theory of Appearance may have its weaknesses but it has the advantage that it allows us to understand how God or Ultimate Reality might be perceived, since we do not analyse the perception. Hence, for an experiencer to perceive God or Ulti-

33 Alston (1991) 55–59.
34 Alston (1991) 55.
35 Alston (1991) 55.
36 Alston (1991) 55. In this way it differs from naive realism.
37 KJV (1984) Exodus 3:2.

mate Reality is simply for God or Ultimate Reality to appear to the experiencer as so-and-so, in the same sense that the roses appear to the person.[38]

2.2.1 Ways of Religious Experiencing

The question is whether the six different ways of experiencing designated earlier are also relevant for religious experiences. Before repeating the test with examples from religious life, we have to know that religious experiences may occur spontaneously but also by way of meditation and prayer or ritual (among others). A spontaneous religious experience is an experience for which no prior preparations were done while religious meditation and prayer are typical systematic disciplines with the aim of obtaining religious experiences. The Catholic mystical tradition makes a distinction between what they call "'acquired' prayer and 'infused contemplation'".[39] The former is similar to the meditation technique known as the via positiva meditation technique or active meditation, in which the meditator focuses on a chosen image. The image may be an inner image or an image situated somewhere in the room or place where the meditation takes place. Acquired prayer and active or via positiva meditation have in common that the meditator uses his or her mental or linguistic abilities, i.e., the person is active during meditation. However, besides concentrating on a picture or other object of meditation acquired prayer may also include, praying, and praising God as in Taizé masses where a short and simple Christian song is repeated many times.

In infused contemplation, on the other hand, the person remains completely passive. There are no mental or linguistic efforts made. This is similar to the passive or via negativa meditation technique, which roughly, is a meditation technique in which the meditator concentrates on emptying his or her mind of all thought. It is said that God or Ultimate Reality is *presented* to their awareness in a way that does not depend in any way on their own efforts.[40] Furthermore, there seems to be a difference in the experience itself when it is obtained through meditation or prayer or by infused contemplation, according to mystics who have experienced both ways.[41]

---

38 I propose that we accept the Theory of Appearance in order to concentrate on the purpose of the current research project instead of on possible bases for an experience of God or Ultimate Reality. Investigating the possibility of perceptions of God would be too immense a task and I am afraid that there is no room for such an investigation in the present project.

39 Alston (1991) 15.

40 Alston (1991) 16.

41 We shall see that there even seems to be a difference between the neural activity associated with the via positiva and the via negativa.

Another meditation technique, know as Transcendental Meditation (TM) founded by Maharishi Mahesh Yogi, is an effective meditation technique used for gaining deep relaxation, eliminating stress, promoting health, etc. The result may be a religious experience but that is not the main goal.[42]

Now we are ready to present the six ways in which a person may have religious experiences. An example of the *participating perceptive sensory way* of experiencing, would be a Christian person who has a religious experience when reading the Bible, because (1) the experiencer participates in the experience, because she or he actively reads the Bible, she or he does not merely witness her or his own reading act; (2) the experience is perceptive, because of the objects involved (Bible texts and pictures) and (3) the experience is sensory, because it needs some of the five senses for reading purposes.

An instance of the *participating introspective sensory way* of experiencing would be a religious experience obtained through a meditation technique in which the meditator focuses on a mantra (i.e., a sound made by the meditators themselves and which they constantly repeat), because (1) the experiencer participates in her or his meditation, i.e., she or he makes the sound and concentrates on it; (2) the experience is introspective as there are no objects in the experience of concentrating on a sound and (3) it is sensory, because the meditator needs some of her or his five senses to listen and concentrate on the sound (at least at the beginning of the meditation).[43]

If we replace the experience of mentally solving the mathematical problem with the experiences obtained through the via negativa meditation technique we would have an example of a religious experience of the *participating introspective non-sensory way* for the same reasons as mentioned in the experience of solving a mathematical problem. The experience is (1) participating, because to concentrate on something (whether the focus of concentration is an image, a sound, one's breath or emptying one's mind of all thought) needs one's participation, (2) introspective, because the idea of the via negativa meditation technique is to empty one's mind of all thought and (3) non-sensory for the same reason.

An illustration of the *witnessing perceptive sensory way* of religious experiencing would be the witnessing (but not participating in) a mass in a specific church. The experience is (1) witnessing, because the person does

---

42 Meditation may thus be practised for other than religious reasons. However, since the project concerns religious experiences, I shall only consider religious meditation and the associated experiences.

43 We shall look at the process of meditation in more detail later.

not participate in the mass (the liturgy, singing, etc) but merely sits and watches, (2) clearly perceptive and (3) sensory.[44]

The *witnessing introspective sensory way* of experiencing, is probably the best known way of having a religious experience, namely, experiencing *God* in nature.[45] Take, for instance, experiencing God's beauty in the beauty of a cold winter's day in the forest, when one of the effects of sunlight and severe cold is that it seems as though thousands of diamonds twinkle on the snow-clad branches of the trees and on the snowy carpet you walk on. The experience is had in a (1) witnessing way, because one does not participate in the experience, one apprehends God's beauty, (2) introspective way, because God's beauty is not an object (in the Yandell sense) and (3) sensory way, because one has to use some of the five senses to experience God's beauty: God's beauty is everywhere in the winter landscape.

Finally, if we change the witnessing of the mass from the experiencer attending it in a church to him or her remembering the ceremony on a later occasion, it follows that he or she will experience that particular remembering in a *witnessing introspective non-sensory way*. The experience is (1) witnessing, because the experiencer does not participate in his or her memory, (2) introspective, because there are no objects in the memory and (3) non-sensory, because none of the five senses are needed to remember something. Thus, religious experiences may also be experienced in (at least) the six ways of experiencing designated earlier.

2.2.2 Types of Religious Experience

William James would not have written *Varieties of Religious Experiences* if it were the case that a religious experience is the same for everyone and at all times. Clearly, there is a tremendous variety of religious experience. Alston for instance, writes that "[i]n the widest sense, 'religious experience' ranges over any experiences one has in connection with one's religious life, including any joys, fears, or longings one has in a religious context".[46] To name some: Feeling thankful towards God or Ultimate Reality; feeling guided by God or Ultimate Reality; seeing God or Ultimate Reality in nature; feeling a sense of ecstasy; feeling as if one is outside what Fraser

---

44 However, we should note that there is a grey zone between the experiences one is participating in and those one is witnessing. Praying or singing together with the priest and other people would be to participate even if one watches the mass on television. However, in our example, the person merely witnesses what happens during a mass and does not participate in any way.

45 At least in Sweden.

46 Alston (1991) 34.

Watts called the "skin-encapsulated [individual] ego";[47] experiencing pure consciousness; experiencing universal love; having visions or revelations; experiencing a union with God or Ultimate Reality; etc. Needless to say, it is very difficult to account for all experiences within the framework of this project.

A religious experience is not, as some might believe, the same as a religious insight. However, the latter may, and often does, follow from the former. Moreover, this is not only the case for religious experiences. For instance, Paula's experience of pneumonia after standing in the cold rain for hours waiting for her parents to come home because she lost her key, will certainly give her the insight to take better care of her belongings.

Richard Swinburne suggested five main types of religious experience which he classified from the perspective of the experiencer and which he considers to be mutually exclusive and totally exhaustive of the possibilities. I believe though that one type of religious experience should be added, which according to me, is not covered by Swinburne's classification as will become clear below.

The first type designated by Swinburne is the experience of God or Ultimate Reality which the experiencer perceives through a "perfectly ordinary non-religious object".[48] Among other possibilities, this might include experiencing God or Ultimate Reality in a beautiful sunset or in a piece of Gregorian music. The sunset and the music are the channels through which God or Ultimate Reality is experienced. This type of religious experience may be experienced in a participating perceptive sensory way (reading or listening to music) but also in a witnessing introspective sensory way (experiencing God in nature). Furthermore, the experience may occur spontaneously but also by way of meditation and prayer. The following example expressed by one of my informants illustrates a religious experience of the first type experienced in a witnessing introspective sensory way. "At a given moment, while I was walking in the forest, I looked up and saw an opening towards heaven. I experienced God's peace and returned to this place many times later on."[49]

The second type of religious experience defined by Swinburne is the experience of God or Ultimate Reality which the experiencer perceives through "very unusual public objects".[50] Swinburne gives the example of the *burning bush* which Moses claimed to be God appearing to him in a fire out of that bush. Moses should have experienced God in a participating perceptive sensory way, because:

47 Braud (2003) 124.

48 Swinburne (1979) 250.

49 From one of my informants, my free translation from Swedish, cited with permission.

50 Swinburne (1979) 250.

There the angel of the LORD appeared to him in flames of fire from within a bush. Moses saw that though the bush was on fire it did not burn up. So Moses thought, 'I will go over and see this strange sight – why the bush does not burn up.' When the LORD saw that he had gone over to look, God called to him from within the bush, 'Moses, Moses!'[51]

This type of religious experience is typically a spontaneous religious experience, though it is not entirely impossible that such an experience occurs during meditation.

The third type of religious experience defined by Swinburne is the experience of God or Ultimate Reality mediated through "certain sensations private to himself [or herself], sensations of a kind describable by normal vocabulary used for describing the sensations which result from the use of our five senses".[52] Good examples are, experiencing God in a dream, in a vision, or during a trance. The experiencer having a dream about God then experiences God in a witnessing introspective non-sensory way. Dreams and visions may occur spontaneously but also during meditation. Consider the following experience:

Joy Kinsey [...] c[a]me to the altar for prayer during the service [...] Her intention was "just to kneel and pray and just really totally surrender my will to God for whatever purpose." A minister came over to pray with her, and when he touched the back of her head in a gesture of blessing, she fell backward and lost consciousness. For three hours Joy was unaware of anyone or anything around her. She had the sense of being in a place where a temple was surrounded by a courtyard. The temple had three domed parts to it, attached together so that they formed one continuous building. She began to walk through it, each part beautiful beyond description, but when she came to the threshold of the third part she stopped, for she felt unworthy to enter. As she looked in she saw Jesus sitting on a throne about fifteen feet away, but sitting sideways in relation to her and partially obscured by a lattice. He looked pleased at her having come so far, such as a parent might look upon his or her child taking its first steps. He appeared average in size, solid in appearance, much as she had pictured him.[53]

A specific variant of this type of religious experience could be experiences of God or Ultimate Reality in near-death and life-beyond-life experiences.[54] For instance, my grandmother on my father's side told us about her near-death experience. As she was a Catholic she had received the last anointment with oil, which means that she was dying. She said later when she had unexpectedly recovered:

---

51 KJV (1984) Exodus 3:2–4.

52 Swinburne (1979) 250.

53 Wiebe (1997) 42.

54 The difference between near-death and life-beyond-life experiences is that during the former experience the person is about to die. During the latter experience, the person has died (in a clinical sense) but has been brought to life again by way of heart massage after an accident or operation or other.

I was flying through a long dark tunnel. I saw a light becoming bigger and bigger and I was flying towards it. When I was near enough I saw Him, our Lord. I felt so peaceful and happy. He is so beautiful. Then I realised that Carlos [her youngest son] would become an orphan if I stayed, I was worried and asked to be allowed to return. They did not let me at first but I begged and begged and – here I am.[55]

I propose that my grandmother experienced this in a participating introspective non-sensory way as described above.

The fourth type of experience, then, is described as an experience of God or Ultimate Reality that is mediated through private sensations "yet these are not a kind describable by normal vocabulary".[56] An example of such a religious experience, experienced in a witnessing introspective sensory way, comes from Saint Teresa of Avila who was completely aware of Jesus sitting next to her, even though she did not see him. She was totally convinced that he was sitting next to her and talking to her:

One day when I was at prayer – I saw Christ at my side – or, to put it better, I was conscious of Him, for I saw nothing with the eyes of the body or the eyes of the soul [the imagination]. He seemed quite close to me and I saw that it was He. As I thought, He was speaking to me. [...] All the time Jesus Christ seemed to be at my side, but as this was not an imaginary vision I could not see in what form. But I most clearly felt that He was all the time on my right, and was a witness of everything that I was doing. [...] If one says that one is like a person in the dark who cannot see someone though he is beside him, or that one is like somebody who is blind, it is not right. There is some similarity here, but not much, because a person in the dark can perceive with the other senses, or hear his neighbor speak or move, or can touch him. Here this is not so, nor is there any feeling of darkness. On the contrary, He appears to the soul by a knowledge brighter than the sun. I do not mean that any sun is seen, or any brightness, but there is a light, which though unseen, illumines the understanding.[57]

These experiences can also occur spontaneously and by way of meditation, prayer or rituals such as mystical experiences.

The fifth and last type of religious experience defined by Swinburne is an experience of God or Ultimate Reality which the experiencer "does not have by having sensations".[58] For example, this could be experiencing God or Ultimate Reality intuitively and immediately. One *intuitively knows* that it is God or Ultimate Reality that one experiences. They are typically experienced in a witnessing introspective non-sensory way and are those of

55 From my own memory.
56 Swinburne (1979) 251.
57 St. Teresa (1957) chap. 27, 187–189, replicated by Alston (1991) 13, Peterson et al. (2003) 18. A good friend of mine had a very similar experience of Jesus Christ when sitting on a bench at a market place in Prague.
58 Swinburne (1979) 251.

which it is said that they are ineffable. Such experiences occur both spontaneously and by way of meditation. Nicholas of Cusa maintained:

But I perceive, not with my flesh eyes, which look on this icon of Thee, but with the eyes of my mind and understanding, the invisible truth of Thy face, which therein is signified, under a shadow and limitation. Thy true face is freed from any limitation, it hath neither quantity nor quality, nor is it of time or place, for it is the Absolute Form, the Face of faces.[59]

The extra type I would like to add to Swinburne's list of religious experiences is defined by Phillip Wiebe and can be described as experiences of God or Ultimate Reality with "observable effects".[60] These are experiences in which more than one person is affected. It may happen that two or more persons take part in the same religious experience, that they really *saw* the same thing, or that one person undergoes the experience while another (or others) felt a tactile sensation. Even if it might be argued that Swinburne's types of religious experience also count as experiences where more than one person is involved, I still believe that Wiebe's type is not included. For instance, this type of religious experience does not have a place in Swinburne's first type, because the God experience as understood in Wiebe's type is not communicated through a common public sensory object such as a sunset or a piece of music. Neither does it belong to the second of Swinburne's types, because the experience is not mediated through an unusual, public sensory object such as a burning bush. Swinburne does not consider it in the third and fourth types either, because the experience is not a matter of private sensations, and finally, since the experience is not an intuitive experience of God it cannot be classified as belonging to Swinburne's last type. An example of such an experience of God, which has observable effects, is found in the Bible when Jesus appeared before Peter, James and his brother John at the same time and in the same place.[61] However, the following example is more current and illustrates that John experienced Jesus in a witnessing perceptive sensory way while Nathan experienced Him in a witnessing introspective sensory way.

John shared his room with Nathan. [...] During November of that year Nathan came down with a virus and stayed in bed to recover. [...] John felt sympathy for him and brought him food from the cafeteria when he could, and prayed with him before retiring for the night. As he was praying for Nathan one night he opened his eyes to look at his friend lying about eight feet away. John was shocked to see someone standing over Nathan's bed, but facing and looking at him. John immediately identified the person as Jesus, in part, because of the sense of awe that the appearance of

59 Nicholas of Cusa, *The Vision of God*, cited by Peterson et al. (2003) 19.
60 Wiebe (1997) 74–82.
61 KJV (1984) Matthew 17:1–8.

the person evoked. John was about to tell his sick friend what he was seeing when Jesus reached over and placed his hand on Nathan's forehead and disappeared. At that instant Nathan leaped out of bed and ran down the halls of the dormitory shouting, "I've been healed, I've been healed." Nathan later said that although he did not see anyone, he felt something touch his head. John himself intended to go over to touch Jesus in order to establish his reality for himself, but did not get a chance to do so.[62]

To round off, six possible types of religious experience were presented: (1) Experience of God or Ultimate Reality mediated through a perfectly ordinary non-religious object. (2) The experience of God or Ultimate Reality mediated through very unusual public objects. (3) The experience of God or Ultimate Reality mediated through private sensations of a kind describable by normal vocabulary as used to describe sensory sensations. (4) The experience of God or Ultimate Reality mediated through private sensation that cannot be described by normal vocabulary. (5) The experience of God or Ultimate Reality which the experiencer does not have by having sensations. Finally, (6) religious experiences with observable effects. We also saw that these types of religious experience can be experienced in different ways in accordance with those designated above, and that they may occur spontaneously and by way of meditation, prayer and other techniques aiming at experiencing God or Ultimate Reality.

2.2.3 Non-Religious and Religious Experiences

As mentioned earlier, we should also consider non-religious experiences and how non-religious and religious experiences are related to human experiences. A non-religious experience is simply an experience without religious elements or a de facto religious experience that is not accepted as such by the experiencer. We saw that a human experience in general is an experience of things and events whose existence ultimately is constituted of properties and relations, actions and interactions of whatever the physical and psychological worlds treat of and is an expression of a relation between the experiencer and the experienced. Furthermore, all experiences are experiences of an aspect of the subject of experience and sometimes even of an object and everything we experience is presented to us in a certain way. Accepting these conditions has the advantage that both religious and non-religious experiences can be regarded as *interpretations* of human experiences. Religious experiences, like all experiences, are constituted of properties and relations, actions and interactions of whatever the physical and psychological worlds treat of. My view, however, should not be confused

62 Wiebe (1997) 76.

with the idea that religious experience is a *dimension* or *aspect* of human experiences.[63] This is not what I have in mind since this is the idea that

> [w]e are oriented towards God. [That t]his original experience is always present. [...] This unthematic and ever-present experience, this knowledge of God which we always have even when we are thinking of and concerned with anything but God, is the permanent ground from out of which that thematic knowledge of God emerges which we have in explicitly religious activity and in philosophical reflection.[64]

The idea is that a human being's primary endeavour is to experience God. This is why there is an aspect or dimension of religious experience in all human experiences. This is not what I intended here.

Are there no differences then between non-religious and religious experiences? Alston suggests the following differences. One is that religious experiences are not shared by all human beings.[65] However, this is not a very good suggestion. Seeing, for example, is an experience that is not shared by all human beings. Another difference that Alston presents is that "God is a rare phenomenon except for a very few souls".[66] Unfortunately, this is not a successful proposal either, because there are many phenomena that are rare for most people. An instance of such a phenomenon is being hit by lightning. A third difference he names is that sense perception normally is very colourful, rich, vivid etc. in contrast to the experience of God or Ultimate Reality, which according to him, is "dim, meager, and obscure".[67] This difference does not really hold either, not if we accept the bright and colourful narratives of mystics such as Teresa of Avila and other people describing their experience of God or Ultimate Reality. Alston's differences are, in my view, not really eloquent.

A significant difference, I believe, lies in the context of the experience, namely, in the fact that the context of the religious experience (or what the context of the religious experience is a part of or related to) is mostly (but not always) presupposed by the experiencer. In other words, there may be a specific commitment between the experiencer and the context of his or her religious experience (or what the context of the experience is a part of or related to) in the form of belief, wonder, seeking, doubting, will to understand, etc. There is no such commitment between me and the computer I experience writing on or between me and the sun shining through my window.[68]

---

63 Alston (1991) 34.
64 Rahner, as cited by Alston (1991) 34.
65 Alston (1991) 36.
66 Alston (1991) 36.
67 Alston (1991) 36.
68 Of course, experiencing bungy jumping may have something to do with wondering, seeking or doubting such as seeking one's own limits, but that is not what is meant here.

2.3 Psychical Experiences

There is yet one more type of experience that should be regarded for our purpose, namely, psychical experience. As mentioned in the introduction of this chapter, the reason for making a distinction between religious and psychical experiences is, because as we shall see below, not all neuroscientists pay attention to the differences between these experiences and some classify all psychical experiences as religious ones. One difference, according to William Braud, is that psychical experiences (but not religious experiences) are those experiences that people use to influence the world around them using other than the common five senses.[69] This influence need not be understood in a negative sense. People who use their special abilities often use them to help other people. These experiences may occur spontaneously and by way of meditation, trance or rituals. Experiences that are usually classified as psychical experiences are clairvoyance, precognition, telepathy, and psychokinesis.[70] The borderline between religious and psychical experiences is not always very clear, i.e., religious experiences may also be manifested by way of, for example, clairvoyance (which in Christian traditions probably would be regarded as a vision) and precognition (which in Christian traditions probably would be considered as prophecy)

The first one, *clairvoyance*, can roughly be described as having a sudden knowledge of some objective event, usually at a distance.[71] Clairvoyance is a way of mind-to-object communication or interaction. Take the example of the old lady, let us call her Betty, who told the postman when he delivered a package to her one morning, that he had to go home immediately because something had happened to his wife. The postman did as he was told and found his wife unconscious at the bottom of the stairs. In my view, such an experience could even be a religious experience. For instance, the woman might be a religious person for whom it would be natural to attribute her experience to the foresight of God or Ultimate Reality while seeing herself as a servant of God. Such an experience could then be in line with the doctrines and soteriology of the religious tradition in question.

Precognition is having accurate knowledge of future events that have not yet occurred and that could not have been predicted on the basis of rational inference nor accounted for on the basis of chance coincidence.[72] A good example is the narrative of a young woman, let us call her Mary, who had a night out with her husband. Suddenly, while they were watching a movie at the cinema, she smelled fire. She told her husband, they started to look

69 Braud (2003) 124.
70 Braud (2003) 124.
71 Braud (2003) 124.
72 Braud (2003) 124–125.

around, but everything seemed to be in order. After a while though, the smell of fire became unmistakable. Mary ran out, stopped a taxi and drove home as fast as possible. Entering the living room she saw the baby-sitter sleeping on the couch. She ran to the kitchen where her attention was drawn to the shining red overheated plate of the electric cooker. The little baby was sleeping upstairs. In my view, this experience could also be a religious one for similar reasons as given above.

The third experience, a *telepathic experience*, is a way of distant mind-to-mind communication or interaction.[73] This psychical experience is probably known to most of us even if we do not think about it in such terms, namely, when we are deeply in love and words are not necessary to communicate our feelings to each other, nor is our presence. Telepathic experiences may also be classified as religious ones if they fulfil the two criteria mentioned above.

Finally, *psychokinesis or telekinesis* is known as mind-matter interaction in which one's attention and intention may directly influence other persons, other living systems or inanimate physical systems usually from a distance.[74] The most well known person having such abilities is probably the Israeli Uri Geller. He demonstrated his abilities in the Seventies in public and bent metal spoons and keys. He even managed to manifest these effects over great distances and in multiple locations.[75] Another example would be Carrie, the fictive character in Stephen King's novel by the same name, who, by way of telekinetic powers, burned down her school. Should this experience be characterized as religious? In my view it is more difficult to see a religious variant. There may be cases in which people maintain that God provided them with such psychic gifts but these gifts are not typically associated with religious experiences nor with the doctrines and soteriology of the religious traditions.[76]

Hence, it seems to be important to pay attention to how people describe their psychical experiences not least when they consider them to be religious. We have to investigate the way in, which according to them, *things* (objects and aspects of subjects) appeared to them. We also have to consider whether these experiences can be classified as doctrinally and soteriologically belonging to the religions.

---

73 Braud (2003) 124.

74 Braud (2003) 125.

75 Tests on psychical experiences are performed, among other universities, at the Princeton Engineering Anomalies Laboratory (PEAR) at Princeton University by Mario Varvoglis, Ph.D.

76 Not all people who maintain having such a psychical gift really have the gift but that is not att issue here.

2.4 Explaining Things and Events

It is one thing to experience, but we also want to explain our experiences. The question we have to answer now is what does it mean to explain something? When we explain something, our purpose is always to clarify that something, to clarify a specific item of that something, to make that something easier to grasp. Therefore, it is not unusual that the same issue is explained in different ways in order to reach different types of readers (children, adults, teachers, academics, professionals, lay-men etc.).

2.4.1 General Explanations

How do we commonly explain things and events? Generally we understand by explaining something, making something intelligible or understandable. We explain something by way of showing why or how something occurred. We quickly understand that just about anything can become a victim for explanation when we listen to all the *why* questions of a six-year old child. Mostly, explanations to *why* questions start with a *because*. In other words, we usually explain events and things by way of their *causes*. However, there can be many causal factors why something occurred. For example, the reason why I feel hungry, i.e., the cause of my feeling hungry, may be that I have not eaten since breakfast which was eight hours ago, that I have a high metabolism, or that I have some stomach disease that gives me the feeling that I am hungry even if I am not. Thus the cause depends on *contextual* considerations.

Another important explanatory form is the *how* form. *How* explanations are *descriptive*. They explain the process of an event or the cause of a thing or being. Often, how-explanations involve causal explanations. For instance, take the question *how* did you come to work this morning? Well, I came by car (Process of driving the car from my home to my work), *because* I missed the bus and had an appointment at nine so I could not wait for the next one (missing the bus and having an appointment caused me to take my car).

When it comes to human actions, we want to have a *rational* or *intentional* item in the explanation. People do not just do things; they mostly have a *reason* for why they do things. A fish eats the food that you pour in the aquarium, because eating is one of the instincts of a hungry fish. My eating many vegetables has two reasons, one is, because I find vegetables very delicious (desire). The other reason is that I disagree with the way animals are treated (ideology). However, I know that I need proteins and shall therefore eat chicken now and then and expand my meals with different sorts of beans (rational).

We also explain things by way of *categorizing* them. We know, for example, that the difference between a dolphin and a swordfish is that the former belongs to the zoological category of mammals while the latter belongs to the group of fish. But they have in common that both swim in the oceans and nourish themselves on small fish and plankton. Thus, the dolphin and the swordfish are classified, put in their accurate classes.

Other types of explanations are explanation by *identification*: for example, water is equal to H2O. In other words, we explain something by establishing its identity. We also explain things and events by referring to *analogies* (God is like a father), in terms of *understanding* and probably by way of many other factors.

2.4.2 Scientific Explanations

In the context of scientific explanations (as in other academic contexts) we need to distinguish between real and spurious understanding. Actually, the way scientists explain things and events is not really unlike the way we generally do, but the expectations are different. One expects the explanation to be *objective*, in the sense that science has established.

However, because scientific explanations are said to be objective, it does not imply that there is no subjectivity involved. This became obvious when science was confronted with quantum mechanics. The late Per-Olov Löwdin said that "no measurements can be done without an observer being around, and this observer then cannot be separated from the experiment (measurement)".[77] What Löwdin means is that using quantum theory, one can only interpret the universe "through its projection on the observer's consciousness by way of his or her senses".[78] How then could such subjectivity be considered as "physical reality" and "scientific objectivity"?[79] Löwdin answers:

> Actually, one quietly departs from the assumption that there is a universe out there of which the characteristics can be described, because of its interaction with our senses. The fact that vast numbers of physical observers experience these projections in a similar way is interpreted as depending on two circumstances: a) Besides the observers themselves, the outer universe is the same for all observers (assumption), and b) the universe is comprehended similarly by all observers due to them having a similar biological construction (observation).[80]

---

77 Löwdin, as quoted by Holte (1983) 56, my translation from Swedish. Per Löwdin founded the department of Quantum Mechanics in Uppsala, Sweden.

78 Holte (1983) 56, my translation from Swedish.

79 Holte (1983) 56, my translation from Swedish.

80 Holte (1983) 56–57, my translation from Swedish.

Thus, in my opinion, objectivity should rather be understood as *intersubjectivity*.

Usually, in science things and events are explained by way of models. Unfortunately, there is no standard definition of a scientific model. Generally, one could say that a model represents something different than itself.[81] A model can be defined as "a mental construction that based on the reality, reproduces the main components and relationships of the analyzed segment of the reality".[82] Similarly, to Ulf Görman, models have intellectual relations with reality. He maintains that they serve as tools for understanding reality. These models should not be regarded as entirely realistic, but as simplified versions of the reality to be explained and thus always in some sense selective.[83]

According to José Maria Cavero and Esperanza Marcos, this comprehension of models has been used in empirical science since Galileo, because scientific models are constructed to be able to study some sequence of reality. Thus, according to Cavero and Marcos, these models (there are others) must be "*isomorphic representations* of reality, simpler than reality and emphasizing its main features or, at least, those most relevant to the object of research".[84] However, similar to the fact that there is no consensus concerning the meaning of model, there is no consensus concerning the meaning of isomorphism. Zbigniew William Wolkowski says that isomorphism may be defined in three different ways: Firstly, as "a one-to-one correspondence between elements in different systems such that the relationship between the elements is preserved"; [85] secondly, as "a perfect analogy", [86] i.e., everything that the one system has, the other also has and whatever happens in the one system, also happens in the other and conversely; thirdly, "whenever the symbols of one mathematical model stand in one-to-one correspondence with those of another, including the symbols of relation, and whenever the relation is preserved when a pair of symbols in one model is replaced by their counterparts in the other, the two models are isomorphic".[87]

The philosopher Ronald Giere does not understand *systems* in terms of *models*. He writes that "[instantial] models will [...] be isomorphic, that is, there will be a one-to-one correspondence between the elements of the different models".[88] However, the models in question must be *physical* and not

81 Görman (2002) 1374.
82 Pelayo, as quoted by Cavero/Marcos (2004) 1.
83 Görman (2002) 1374–1375.
84 Cavero/Marcos (2004).
85 Wolkowski (1993) 61.
86 Wolkowski (1993) 61.
87 Wolkowski (1993) 61.
88 Giere (1999) 42.

merely mathematical.[89] Cavero and Marcos understand models as used in empirical science as *isomorphic representations* while Giere makes a distinction between *interpretive or instantial models*. Instantial models provide an interpretation of a set of uninterpreted axioms and *representational models* are “tools for representing the world”.[90] Görman disagrees that models are isomorphic to reality but makes a distinction between models that (simplified and selective) *represent* reality, for example, simplified models of the universe; models that *replace* reality, for example, rats (instead of human beings) in medical research and *analogies*, for example, waves of water as analogies for waves of light.[91]

Clearly, there is no real consensus on the understanding of isomorphism. Nevertheless, there might be a way to classify the different notions of isomorphism that could facilitate understanding. While discussing the issue at the department of Theology in Uppsala, Eberhard Herrmann put forward the idea of distinguishing isomorphic models by dividing them into *representative* isomorphic models and *structural* isomorphic models. From what is said above it follows that representative isomorphic models should correspond one-to-one with the reality or substructure of that reality which it is meant to represent in such a way that the relationship is preserved.

A good example would be building models for *archaeological ships*. The constructors use precise details (material such as different sorts of wood, paintings, sizes of different parts, etc.) derived from the archaeological remains and preserved documentation of the ship to build the model to scale. Such models are used to rebuild the ship in full size, to put in a historical museum or as research and study objects.

A good illustration of an isomorphic structural model is a city map. The different lines on the map represent the streets, railway lines, water, parks and squares of the city. However, these lines (and thus the map) are not representative in the sense that they are reconstructed with the same details as the streets, railway lines, and so on, of the city. Structural models are also scale models but they do not correspond one-to-one with the *real* city in the way the scale model of the ship corresponds with the *real* ship.

However, city maps are based on experience, and at least nowadays, on measurements. For instance, we have the possibility to take satelite pictures, a possibility Gerardus Mercator did not have. Hence, the exactness of city maps, (and other structural isomorphic models) varies with the amount and quality of information scientists have at their disposal.

89 Giere (1999) 42.
90 Giere (1999) 41–43.
91 Görman (2002) 1374–1375.

Most isomorphic models used in science, I would say, are structural isomorphic models.[92]

However, not all models applied in science are understood to be isomorphic to reality. According to Cavero, some disciplines connect "models with analogies and metaphors, and [some define] scientific theories as mathematical models".[93] The latter are called "*theoretical models*" by Giere.[94] Clear examples of such theoretical models are Newton's Law, The Schrödinger Equation and the Principle of Relativity. However, I believe that the Deductive-Nomological Model (or DN) may also be ranked among the theoretical models.

The Deductive-Nomological explanatory model (DN), created by Carl Hempel and Paul Oppenheim explains events and things by way of scientific laws and theories.[95] It is an idealised model that favours simplicity and aims to replace dubious and "ambiguous talk about the world with clear and precise talk about language.[96] The DN model consists of an *explanandum* (a description of the phenomenon to be explained), and an *explanans* (a set of sentences being offered to account for it such as a scientific law). Thus, it is sentences that are explained rather than events. The explanation is deductive and ignores contextual factors.[97]

Another model used frequently in science is the causational explanatory model.[98] That something x caused something y implies that x precedes y. Furthermore, if something x is true, then something y will follow, is logically valid. That the eight o clock train was delayed (x) caused me to miss my appointment with my son (y). However, it is important to note that causal necessity (in the causal model) is not equal to logical necessity, because causal necessity (in the causal model) only refers to a relevant causal factor that is specific for a thing to be in a certain way or for an event to happen. To illustrate this, if I hit the vase with the hammer, the vase will break, i.e., my hitting the vase with a hammer causes the vase to break. According to the rules of logic this implies that if the vase is not broken, I did not hit it with a hammer. This clearly is not what is meant by causal necessity as used in scientific causal models. Rather, what could be meant is that the hammer

92 I have presented different (but not all) models that can be used in science but nothing about how scientists interpret their models. Some scientists may regard their structural isomorphic models as real representations of the world and others may consider their representative isomorphic models as structural. However, analysing this further would lead me away from the purpose of the present chapter.

93 Cavero/Marcos (2004) 1.

94 Giere (1999).

95 in 1948.

96 Hobbs (1993) 5.

97 Hobbs (1993) 3–10.

98 Hobbs (1993) 13–16.

and the vase are constructed of such material that when the hammer strikes the vase, the vase will break. The causing factor is the smash of the hammer being constructed in a specific way, striking the vase.

A distinction that is interesting to make is the distinction between explanatory and exploratory models. Explanatory models are models as defined above which are used to explain that which is already known, for example a model of the biological functions of the human heart. Exploratory models, however, may lead to further knowledge and insight, such as a model explaining the impact of rapid climate change on human behaviour. This model not only explains expected changes in the climate but also explores the consequences this could have for human behaviour. However, there is no real borderline between the two. They may overlap each other as in the following theoretical explanation: explaining a known quantum state by way of equations also invites new thinking concerning the moment of the collapse.[99]

However, not all explanations are adequate ones. Sometimes explanations are *uninformative* and hence do not provide the information that was asked for even if the explanation is relevant. If I ask why you are angry with me and you answer that the central nervous system is constructed so that it supplies the body with adrenaline in order to be able to defend itself when in danger, you would have given me an adequate, but uninformative explanation, because you would not have answered my question. That was not what I was asking. Rather I wanted to know what I have done to make you angry with me. However, uninformative explanations need not be irrelevant. *Irrelevant explanations* are those which have nothing or very little to do with the subject of enquiry and are often given by persons who do not, or not satisfactorily, possess the necessary knowledge about the subject in question.

According to Newton-Smith, "explanations are sensitive things. Adding a simple truth to a good explanation may destroy it".[100] Actually, what he means is that (even) scientific explanations last only until a counterarguement or some element that makes the explanation invalid, is presented.

2.4.3 Explanations in Theology and Religious Studies

Let me begin with reminding you about the distinction made in the introduction between different understandings of theology. I suggested theology,

---

99 We could ask the question whether an exploratory model could exist without an explicative context. Do we not always depart from the known to discover the novel?

100 Newton-Smith (2000) 128.

in this study, to be understood as non-confessional. The question now is how theology and religious studies explain things and events.

We also saw that the research field of theology, and sometimes also of philosophy of religion, both concern the world and its substructures and the Ultimate World and its substructures. Accordingly, some theologians and some philosophers of religion both study how people interpret the Ultimate World and try to explain the Ultimate World. The research field of religious studies only concerns the world and its substructures and studies how people comprehend the Ultimate World. Some philosophers of religion affiliate with religious studies.

Thus, we may establish that researchers of religious studies explain things and events in a similar way to scientists. For instance, psychologists of religion, in the course of their research, use similar models to those applied within traditional psychology. Thus, within religious studies, the same models are used as within social sciences. The research on both psychology and psychology of religion concerns, among other aspects, the psychological implications of the environment on the individual (or group of individuals). Another example is, since both theology and religious studies perform historical inquiries and do field studies, etc. the models used will be similar to those used in the corresponding traditional discipline.[101]

Philosophy of religion is somewhat special. Some philosophers of religion will only apply models similar to those used in traditional philosophy while others will also apply models similar to those used in theology, i.e., models representing the Ultimate World and its substructures. The latter may be based on different evidence, such as, individual descriptions of, for example, religious experiences, anecdotes, hypotheses, historical facts, and other evidence that fits.[102] A most obvious difference between scientific and theological models is that it is very difficult to design a model to represent the Ultimate World, because if we want to model what we interpret as being our relation to God or Ultimate Reality, and since languages are evolved for natural world applications, we lack natural language predicates to apply to God or Ultimate Reality. Since language, then, only serves to construct analogies, religious explanations naturally involve analogies and metaphors.[103] An example of such an analogy is given by Edward Schoen:

> By knowing that God is that which provides for his needs, Fred can conclude that God is not a rock, a threatening cloud, or a bump in the night on the straightforward

101 Since religious studies consist of several disciplines, the list of the models used could be quite extensive.

102 There are different models such as different realist models (ranging from anti-realist models on the one end to metaphysical realist models on the opposite end) but as the debate concerned with realism lies outside the topic of my project I shall not go further into the matter.

103 Analogy and metaphor are understood as synonyms.

grounds that things such as these are incapable of supplying food, clothing, and shelter.[104]

God is compared with a caring person (father, mother or other carer). The analogy concerns the function of God to care for God's creation.

Thus, theological models about the Ultimate World are based on experiences of one's relation with the world to represent experiences of one's relation with God or Ultimate Reality.

Nevertheless, models based on analogies and metaphors should not be dismissed as inadequate or bad. For instance, as mentioned in the introduction, Barbour maintains that something which both the sciences and theology have in common is that both disciplines need to work with hypotheses, metaphors (figurative language), explanatory models, etc. to construct theories about the world. This means that science also uses analogies for research purposes, something that was even confirmed above. Thus, there are not only similarities in the use of models (methods) between religious studies and science but also between theology and science. Theologian Sven-Erik Brodd and Paleonthologist Lars Werdelin present a specific example of science and theology using similar explanatory tools.[105] They maintain that some branches of the natural sciences such as evolutionary biology and some branches of theology, such as ecclesiology, use scenarios to explain things and events.

To explain something by way of a scenario, as done in evolutionary biology, is a complex issue. One has to add as much information as possible about, for example, function, environment, adaptation and so forth to the information present in the item to that which is to be explained. Brodd and Werdelin compare this complexity with the complexity we meet in theological explanations. A theological scenario may include different pieces of evidence such as individual descriptions, anecdotes, hypotheses, historical facts, and other evidence that fits.

Brodd and Werdelin assert that scenarios and sometimes even their hypotheses cannot be falsified.[106] What do they mean? Assume that from the inductive given data that all sheep are black, we put forward the hypothesis that these sheep are family. Assume further that we, by way of this hypothesis, deduce the following evolutionary scenarios: (1) Both parents were black for all of them. (2) One black gene was dominant and caused all sheep to be black. (3) The environment is such that all sheep are black.

---

104 Schoen (1985) as cited by Hobbs (1993) 194.

105 Brodd/Werdelin: A paper they presented at the conference "Success and Failure – Bridging the Interdisciplinary, in Umeå, Sweden, May 2004.

106 Brodd/Werdelin (2004).

Can the hypothesis be falsified? It could be argued that the hypothesis that all sheep are family does not logically follow from the scenario that the environment is such that all sheep are black. However, this is not sufficient, because replacing the hypothesis, for example, with all sheep belong to the same species, gives us the same variety of environmental scenario namely: (1) Both parents were black for all of them. (2) One black gene was dominant and causes all the sheep to be black. (3) The environment is such that all sheep are black.

If Brodd and Werdelin are right that some scenario models are used in a similar way in both theology and science to explain things and events, then it follows that theological scenarios cannot be falsified and not always the hypothesis either. However, the question could be raised whether the scientists using scenario's in their research still perform science. Clearly, the criteria of simplicity, repeatability and testability which science established are not satisfied.

Are theological models isomorphic representations of Ultimate Reality? Since we do not know how God or Ultimate Reality (or the substructures of the Ultimate World) really look like, it is questionable (but not entirely impossible) whether theological models are *representative* isomorphic models. Because if they were, then theological models of God or Ultimate Reality would have to correspond one-to-one with Ultimate Reality of the Ultimate World. However, it cannot be concluded from this that theological models could not be *structural* isomorphic models of Ultimate Reality. Structural isomorphic models, as said above, are not reconstructed with the same detail as the reality and they are scale models. Understood as such, I believe that theological models might be structural isomorphic, imperfect models of the perfect.[107]

Something that theology, religious studies and science have in common is that they explain things and events both by way of explanatory and exploratory models. A prominent difference, however, between science and theology concerns the phenomena which are the focus of their explanations. Scientists seldom have reasons to doubt the existence of the natural phenomena they are about to explore. Theologians and others studying the Ultimate World, on the other hand, have to be critical towards the reality they study, because explaining the Ultimate World includes more subjective aspects for its explanations than the subjective aspects involved in scientific research.[108] However, to be critical is not equal to denying, rather it leads to careful examination. Even if theological explanations are said to be subjec-

---

107 This issue will be further developed in my present research project named *Con-Science about Consciousness*. For example, it has to be investigated whether and how theological structural isomorphic models of the Ultimate World could be tested on their validity.

108 See Löwdin above.

tive, this need not imply that there could not be objectivity involved. For example, it could be assumed that God or Ultimate Reality exists and is the same for all observers, along with the presupposition that observers (people having religious experiences) apprehend God or Ultimate Reality similarly due to their biological condition. In this case, I believe, the criteria for something to be real would also be a matter of intersubjectivity.

Similar to scientific explanations, theological ones and those presented by the studies of religion, are not always adequate or relevant either. Also these explanations may be uninformative and even irrelevant.

I have not said anything about how things and events are explained in religious contexts. As we saw earlier, the basic difference between theology and religion is that no academic education is needed to practise religion. But when it comes to practising theology, special training and an academic degree are typically required. Accordingly, to explain things and events within the context of religion can be compared with how things and events commonly are explained. The explanations related to religious practice are not expected to satisfy specific academic criteria nor to be objective.

2.4.4 Explaining Experiences of Others

The next questions are, how do we explain the experiences of others, especially religious ones, and how can we understand the experiences of others? When we explain our own experiences we are able to explain them directly, i.e., we explain our own experiences with our own explanatory tools. When we explain the experiences of others we have no choice to explain them other than by our own explanatory tools, but we explain the experiences of others indirectly. Concerning the explanation of religious issues like religious experiences, we have to study carefully beliefs, doctrines and the soteriology available in the particular religion of the person whose experiences we want to explain. We also have to study the person and his or her culture and its rules and practices. This view is shared by Hans Hof. He writes, for example, that the stories in the Bible are based on religious experiences of people who lived in those times. These people tried to communicate their experiences to their environment. In doing so they used the languages and pictures that belonged to their culture and time-period. Hof also says that these reports of religious experiences may, in the long run, become but stories for some people. For others, though, they continue to be testimonies of religious experiences, because according to Hof, these people have, to say it in Mircea Eliade's words, *relived* the experiences.[109] In other

109 Hof (2000).

words, they have had similar experiences and came *as close as they could get* to the experiences of the people of the Bible.

Let me clarify what I mean by way of what Michael Tye has said about pain. According to him, anyone with a complete understanding of the term pain and a perfect awareness of the physical facts of it, can still coherently ask why such-and-such neural activity feels the way it does or why it feels at all. Neurologically, the experience of pain is associated with the cerebral neocortex. However, these physical and functional stories are incomplete. According to Tye, a person born without the capacity to feel pain, and kept alive in a very carefully controlled environment, will not know what it is like to experience pain. In other words, this person will never have the relevant perspective. No matter how much information this person is provided with as to the functions of the brain, its neurosystem as well as its chemical system, he or she will never fully comprehend what the feeling of pain is like. The experience of pain remains but a story. As Tye points out, to be completely understood for what it is essentially in itself, each such phenomenal state *s*, will require the adaptation of one particular point of view or perspective, i.e., the perspective provided by undergoing *s*. Furthermore, this is so whether or not the state of pain is physical or mental, or both. Tye stresses that it is the person introspecting her or his phenomenal states that forms the conception of what it is like to be in that state at that particular time. These persons are then able to testify what pain is like.[110]

Since religious experiences are interpretations of human experiences and since all human experiences are correlated with a neural basis (brain activity), religious experiences also are. Following Tye's example, we could establish that anyone who has a complete understanding of the term religious experience and is fully familiar with the parts of the brain that correlate with religious experiences and/or which bodily reactions are associated with them, can still coherently ask why such-and-such brain processes occur at all.

Let us assume (per impossible) a person who is born without the capacity of having a religious experience but who thoroughly learns everything about religious experiences, doctrinally and soteriologically central to the religions. Even though this person might behave as if he or she has frequently undergone such experience and even if he or she becomes very skilled at pretending, he or she will not know what it is to have a religious experience. The religious experience will be nothing but a story. Religious experience and the experience of pain both have the same phenomenal state *s* so that to be completely comprehended for what it essentially is, the adoption of one particular point of view or perspective, i.e., the perspective

110 Tye (1999).

provided by the experiencer undergoing *s*, is required. However, in my view, such a perspective is not possible to adopt, because experiences, religious or other, are particularly subjective, depending upon our individual neurological, biological and physiological uniqueness. Therefore, I propose that what is required is the adoption of the perspective provided by undergoing an *s-like-experience*. Let us call such a perspective *the closest you can get perspective*. Having such a perspective allows people to testify and communicate with others about what a religious experience is.[111]

We can never understand the experiences of pain or the religious experiences of others as we understand our own, but we need our own experiences of pain and religious experiences to really understand those of others.[112]

2.5 Summary and Conclusions

A human experience is an experience of things and events whose existence ultimately is constituted of properties and relations, actions and interactions of whatever the physical and psychological worlds treat of. It is an expression of a relation between the experiencer and the experienced. To have a human experience is to be in a conscious state in which the person phenomenologically seems to be aware of an aspect of the subject of experience and sometimes, but not necessarily, even of an object relative to the experiencer. The aspect of the subject and object of the experience is always presented to the experiencer in a certain way. Human experiences were divided into those having *de facto evidence* and those having *effective evidence*, i.e., a human experience has de facto evidence for a claim if the experience meets the relevant conditions. A human experience has effective evidence if it has both de facto evidence and is accepted as such by the experiencer. Furthermore, an experience with de facto evidence may be accepted as such and thus count as an experience with effective evidence for a person who did not undergo the experience him- or herself, while the person having the de facto evident experience does not recognise it as such. All human experiences may be experienced in the following ways: (1) The participating perceptive sensory way of experiencing. (2) The participating introspective sensory way of experiencing. (3) The participating introspective non-sensory way of experiencing. (4) The witnessing perceptive sensory way of experiencing. (5) The witnessing introspective sensory way of experiencing.

111 Runehov (2004).

112 Some would argue that it is sufficient to have empathy to understand the experiences of others. According to me, however, similar to theoretical knowledge, empathy would not give us the introspective knowledge needed to know what it is really like to be in such and such a state and to be able to testify about it.

(6) The witnessing introspective non-sensory way of experiencing. It was maintained that religious experiences and non-religious experiences are interpretations of human experiences.

An experience is a religious experience when the following criteria are met. Firstly, when it is an experience in which it seems that God or Ultimate Reality, being the absolute ground of everything that is (and ever will be), appears to a person in a way that is in accordance with the doctrines and soteriology essential to the religious traditions and their continuities. Secondly, when the experiencer recognises the experience as a religious one, i.e., the experience is effective evidence. Assuming that God or Ultimate Reality exists, the Theory of Appearance stating that the notion of God's appearing to the experiencer as so-and-so is fundamental and unanalysable makes testimonies of people saying that they have perceived God or Ultimate Reality plausible. For the experiencer to perceive God is simply for God to appear to the experiencer as so-and-so. Religious experiences can occur spontaneously but also by way of meditation, prayer, rituals, etc. I briefly regarded three meditation techniques: the via negativa, the via positiva and transcendental meditation. It was maintained that the via negativa meditation technique corresponds with the infused contemplation while the via positiva meditation technique is similar to acquired prayer. The transcendental meditation technique is an effective meditation technique for gaining deep relaxation, eliminating stress, promoting health, etc. Six possible types of religious experience were presented: (1) Experience of God or Ultimate Reality mediated through a perfectly ordinary non-religious object. (2) The experience of God or Ultimate Reality mediated through very unusual public objects. (3) The experience of God or Ultimate Reality mediated through private sensations of a kind describable by normal vocabulary as used to describe sensory sensations. (4) The experience of God or Ultimate Reality mediated through private sensation that cannot be described by normal vocabulary. (5) The experience of God or Ultimate Reality which the experiencer does not have by having sensations. Finally, (6) religious experiences with observable effects.

A non-religious experience, then, is a human experience without religious elements or a de facto religious experience of which the religious elements are not recognised as such by the experiencer. It was also said that a significant difference between non-religious and religious experiences is that there may be (allthough not necessarily) a commitment between the experiencer and the context of his or her religious experience whereas this is probably not the case between the experiencer and the context of the non-religious experience.

We also considered the following psychical experiences: clairvoyance, precognition, telepathy and psychokinesis. The borderline between religious and psychical experiences is not sharp, hence it was concluded that it is important to pay attention to how people describe their psychical experiences.

Keeping these accounts concerning religious experiences in mind will help us to explore which religious experiences neuroscientists are able to study by neuroscientific means and which ones they are probably not, given the research methods neuroscientists use today.

When we explain things and events, we do this with the aim of clarifying them, of making them intelligible. Generally we blend possible ways of explaining and we do not really reflect on the plausibility of our explanations. When it comes to academic disciplines however, this attitude cannot be held.

Scientific explanations are maintained to be objective but this does not imply that there is no subjectivity involved. This was illustrated by referring to quantum mechanical research. The subjective aspect of the scientist cannot be separate from the experiment and the objectivity is based on two assumptions: Firstly, that there is a universe out there which is the same for all observers and secondly, that the universe is comprehended similarly by all observers due to the resemblance of their biological construction. Hence, objectivity should rather be understood as collective subjectivity.

Scientists often explain things and events by way of different models. However, there is no standard definition of a scientific model. A model represents something different than itself. Some understand models as isomorphic representations of the world. Since there is no consensus for how this isomorphism should be comprehended and to facilitate the understanding of its different notions, a distinction was made between representative and structural isomorphic models. Representative isomorphic models should correspond one-to-one with the reality or substructure of that reality which it is meant to represent in such a way that the relationship is preserved. However, models are not always understood as isomorphic to reality. They can also be seen as simplified and selective representations of reality, as replacements of reality and as analogies and metaphors to reality or as theoretical interpretations of reality. Scientific models may be explanatory or exploratory. The latter not only explain things and events, they also explore new possibilities. However, it was also said that not all explanations are adequate or relevant. Some are uninformative and others may even be irrelevant.

We saw that the research field of theology and sometimes also of philosophy of religion both concern the world and its substructures and the Ultimate World and its substructures. Accordingly, theologians and some philosophers of religion both study how people interpret the Ultimate World and try to explain the Ultimate World. Hence, the models used by theologians and some philosophers of religion are also meant to represent the Ultimate World and its substructures. A difficulty that theologians encounter when they want to explain the Ultimate World or its substructures is, because languages are evolved for natural world applications, they lack natural language predicates to apply to this Ultimate World and its substructures. One solution is to use analogies and metaphors. Theology and science have in common that they explain things and events both by way of explanatory and exploratory models. One important difference between scientists and theologians is that scientists seldom need to doubt the existence of the natural phenomena they are about to explore. Theologians and others studying religious phenomena, on the other hand, have to be critical towards the Ultimate Reality they study, because explaining Ultimate Reality includes greater subjective aspects for its explanations than the subjective aspects involved in scientific research. I also suggested that because theological explanations are subjective this need not imply that they could not also be objective. Similar to scientists, theologians may at least make two assumptions: God or Ultimate Reality exists and is the same for all observers, and people observing God or Ultimate Reality apprehend God or Ultimate Reality similarly due to their biological resemblance.

Since we do not know what God or Ultimate Reality (or the substructures of the Ultimate World) really looks like, the question was raised whether theological models could be isomorphic representations of Ultimate Reality. If they were representative isomorphic representations, then theological models of God or Ultimate Reality would have to correspond one-to-one with Ultimate Reality of the Ultimate World. Hence, theological models may not be isomorphic representations of the Ultimate Reality in this sense. However, it was suggested that theological models of God or Ultimate Reality may be structural isomorphic models of Ultimate Reality. Structural isomorphic models are not reconstructed with the same detail as the reality and they are scale models. Understood as such, I believe that theological models might very well be structural isomorphic, imperfect models of the perfect.

Nevertheless, not all explanations are adequate or relevant and some may be uninformative and even irrelevant.

Since there is no academic education needed to practise religion, the explanations related to religious practice need not satisfy specific academic criteria nor need they be objective.

To explain the (religious) experiences of others we have to understand the cultural-religious-personal conditions of the person whose experiences we want to explain. We use our own explanatory tools but we use them indirectly. However, to be able to really understand the experience, we need to have a closest-you-can-get (religious) experience ourselves.

As will become clear, these understandings will be helpful in the evaluation of neuroscientific explanations of religious experiences. Furthermore, explanatory and exploratory models will have a central place in the last chapter.

3 Analysing Persinger's Research

Michael Persinger's research, as we will see below, is different from Newberg and d'Aquili's, even if both have performed experimental and non-experimental studies. Newberg and d'Aquili performed a neuroimaging study which they repeated and later extended. Persinger, on the other hand, performed many different experiments. Some of his studies involve clinical equipment such as PET-imaging (Positron Emission Tomography). PET-imaging (or neuroimaging technology) makes it possible to observe parts of the brain that have changed neural activity, compared to baseline, during, for example, (religious) experiences. However, in contrast to Newberg and d'Aquili, Persinger did not use neuroimaging technology to study the neural activity that correlates with religious experiences, for example, those religious experiences obtained through meditation. Instead he studied neural activity associated with schizophrenic hallucinations and epileptic seizure experiences. He maintains that "there is nothing unusual about studying the exception [epileptic seizure experiences] in order to find the rule [the neural activity that generates such religious experiences]".[1] However, mostly Persinger performed neuropsychological studies based on inquiries. This means that the presentation of the analyses and evaluations of the research on religious experience of Persinger and Newberg and d'Aquili will be different. However, I believe them to be comparable.

Persinger's research is meant to give him answers to the following questions: (a) Are religious experiences what they are claimed to be or are they

---

1 Persinger (1987) 17. More about imaging-technology will be presented in connection to Newberg and d'Aquili's neuroimaging research. For the moment, suffice it to know that PET-imaging shows that the themes of experimental phenomena are correlated with the functions of the parts of the brain most metabolically active. For this purpose, the parts of the body that are to be diagnosed are injected with a radioactive tracer. The tracer is a glucose analogue agent, for example, F18-Fluoro-Deoxy-Glucose or FDG. A computer screen shows the metabolically active parts of the brain in colours. It is suggested that the most active parts could be coloured red, the second most active parts yellow, the third most active part green, etc. According to neuroscience, brain activity is considerably reduced during depression, especially in the frontal areas. Consequently, (following the colour range example above) the colour shown in the PET picture would be green with some small yellow, and some blue parts. If the colour turns out to be black, there is no activity at all, i.e., the brain is dead. PET is used in oncology to detect malignancy, because malignant cells demonstrate higher glucose metabolism. PET is used for diagnosing epileptic seizures or epilepsy, Alzheimer and other dementia, schizophrenia, depression, attention deficit disorder, etc.

rather the result of different brain stimuli, including brain damage and illness? (b) If so, can we know which neurological processes generate religious experiences?

A short presentation of Persinger will facilitate the understanding of the intention that Persinger has with his research on religious experience. His studies of religious experience began in the early Seventies. As a professor in physiological psychology, he organised the Behaviour Neuroscience Program (BNP) at Laurentian University in Ontario, Canada, a program which integrates chemical, biological and psychological studies. Persinger is an authorized psychologist, specializing in clinical neuropsychology. As such, he is concerned with people who suffer from mild to moderate brain trauma.

He maintains that the human brain, its microstructure and intricate activity are the source of all human knowledge. Like Newberg and d'Aquili, he is interested in religious behaviour. However, as will become clear, Newberg and d'Aquili's driving force is to understand different types of religious experience. Their interest lies in the religious experiences themselves. Persinger's starting point for investigating religious experiences is to be found in the effects that religious experiences have on human beings and their world. He writes that

> I am concerned about the illusory explanations for human consciousness and the future of human existence. Consequently after writing *Neuropsychological Base of God Beliefs,*[2] I began the systematic application of complex electromagnetic fields to discern the patterns that will induce experiences [sensed presence] that are attributed to the myriad of ego-alien intrusions which range from gods to aliens. The research is not to demean anyone's religious/mystical experience but instead to determine which portions of the brain or its electromagnetic patterns generate the experience. Two thousand years of philosophy have taught us that attempting to prove or disprove realities may never have discrete verbal [linguistic] solutions, because of the limitation of this measurement. The research has been encouraged by the historical fact that most wars and group degradations are coupled implicitly to God beliefs and to the presumption that those who do not believe the same as the experiencer are somehow less human and hence expendable. Although these egocentric propensities may have had adaptive significance, their utility for the species' future may be questionable.[3]

Thus his interest lies in finding the neurological cause(s) of religious experiences.

---

2 Persinger (1987).
3 Persinger (2004).

3.1 God Beliefs and God Experiences

Persinger's psychological study of religious belief contains many interesting issues, but I found it especially interesting to consider what he has to say about the relation between religious beliefs and parental care.

3.1.1 God Beliefs

Let us engage in Persinger's research on God Beliefs which he introduces as follows.

> The neuropsychological basis of God Beliefs is the primary theme of this text. God Beliefs are extraordinary frequent and pervasive behaviors that are displayed in predictable situations. They have determined political strategies and precipitated national conflicts. They have inspired some generations and decimated others. However, the objective understanding of the psychological and neurobiological sources that determine the existence of God Beliefs has been conspicuously avoided. Considering their propensity to emerge during complex and insoluble, personal and public crises, the factors that control God Beliefs should be clearly delineated.[4]

Thus Persinger argues that his study, which is based on neuropsychology, will reveal the *objective bases* of religious belief which have hitherto been avoided. He considers his contribution to be important, because religious beliefs have influenced generations. The aim of his study (as with all of his studies) is to show that religious experiences are illusory and that God or Ultimate Reality cannot exist.

In Persinger *Neuropsychological Bases of God Beliefs,*[5] he divides religious beliefs, or *God beliefs* as he calls them, into two basic components. The first component is the "God Experience",[6] which he calls "Temporal Lobe Transient (TLT)",[7] because according to him, it is altered activity in the temporal lobe that causes the religious experiences. According to Persinger, Temporal Lobe Transients as transients phenomena are loaded with emotional references and are understood as microseizures within the temporal lobe. The second component of God beliefs is the "God Concept",[8] which is determined by verbal conditioning and is associated with different cultures. The God concept is associated with the left temporal lobe which he associates with the analysis and the determination of the perceived objects,

4 Persinger (1987) 1.
5 Persinger (1987).
6 Persinger (1987) 1/16.
7 Persinger (1987) 1/16.
8 Persinger (1987) 1. I am a Christian, a Muslim, a Buddhist, an atheist, I belong to ...

here religious objects. Persinger maintains that "[t]here is great need to understand the biological behavioral basis of God Concepts and the God Experience. The potential long-term negative effects of these behaviors on human survival must be considered now".[9]

Thus, Persinger maintains that God Experiences "occur as small micro-seizures within the temporal lobe."[10] They may involve both positive emotions such as peacefulness and meaningfulness but also negative emotions like anxiety and fear. Furthermore, he refers to contemporary psychological claims that these experiences usually have their starting point in personal crises, because they are triggered by the chemical consequences of stress, "hypoglycemia [changes in blood sugar], fatigue, hypoxia [lack of oxygen], anxiety and a variety of idiosyncratic neurophysiological anomalies [mental disorders]".[11] This list could of course be completed with drugs, such as, mescaline, peyote and LSD derivatives that are said to cause God Experiences. Moreover, music also has the interesting capability of stimulating the temporal lobe and causing a God Experience, as Persinger maintains in one of his studies considered below.[12]

The reason that people have God Experiences is to be found in the evolution of the human race. Persinger maintains that "[a] biological capacity for the God Experience was critical for the survival of the [human] species. Without some experiences that could balance the terror of personal extinction, existence of the human phenomenon called the 'self' could not be maintained".[13] Although it is not known how the very first God Experience was felt, Persinger says that it must have been a kind of "cosmic serenity".[14] The person experienced a kind of quiescence, "there is no fear, no anxiety, only the sensation of becoming a part of everything. [Since] the survival value is real [...] the person can cope with anticipation of self-annihilation".[15] According to Persinger, "[t]he capacity to have the God Experience is a consequence of the human brain's construction. If the temporal lobe had developed in some other way, the God Experience would not have occurred".[16] This biological capacity evolved along with the evolution

9 Persinger (1987) 8.

10 Persinger (1987) 111.

11 Persinger (1987) 111.

12 There are different meditation and praying techniques that include starvation in order to evoke a God Experience. Well known examples are the Christian and Buddhist forty days fasting traditions. Some religious people also climb to the tops of mountains where oxygen becomes thinner, in order to obtain a God Experience. Furthermore, there are yoga techniques that include breathing methods for the same purpose. Concerning drugs we should note that many religions make a clear distinction between the *true* God Experience and those induced by chemicals.

13 Persinger (1987) 12.

14 Persinger (1987) 13.

15 Persinger (1987) 13.

16 Persinger (1987) 13–14.

of the neocortex which is particularly developed in human beings. Religious experiences thus are not only explained in terms of neurology but also in terms of evolution and survival. Thus, when our ancestors faced a danger that was about to destroy them, they simultaneously had a God Experience, because the neocortex supplies such capacity. During this God Experience, their fear disappeared and they were ready to die.[17]

According to Persinger, today, the *terror of personal extinction* is most distinct in personal or psychological crises, the most severe being the anticipation of one's own death, or the death of a loved-one. Persinger agrees with evolutionary psychologists who suggest that psychological crises are the origin of the *God concept*, because the God concept, in most religions, is associated with hope, comfort, and eternal life.[18] When a crisis occurs,

> [and a]s the person loses the strength to struggle [...], fear gives way to a passive resignation. It is then, during the psychological condition of helplessness, that the God Experience typically occurs. No doubt, it may be a consequence of the biological conditions [...] However, the experience can also be predicted by understanding the psychological condition of helplessness. Once this condition occurs, the person feels a sense of expectation that someone will come to the rescue and save the self. [...] Depending on the learning history, the strong expectation develops that *someone will come.* The person begins to read 'scriptures' and suddenly finds another understanding. During the night, a 'soft voice' comes from within the head and says, 'I am the lamb of God' or I am the 'alpha and the omega'. The change that follows is highly variable. Some individuals suddenly return to church, especially if this behavior was learned when they were very young, and find new meaning. Other individuals, without a history of churchgoing become more transcendental or mystical. God has special significance to them. They set aside a part of their house or a portion of their schedule to have a personal communion. [...] There is more to life – an eternal life, with the eternal abstract of God.[19]

Thus, when a person undergoes a deep psychological crisis, it is not that unusual for a spontaneous religious experience to be evoked, because when a person is exhausted or anxious, she or he is more sensitive or open to

---

17 The cerebral neocortex is what makes us human. It is this part of the nervous system that grants us the intelligence that separates us from *other* animals and enables us to create language, art, myth and culture, to make plans and perform complex motor movements. For instance, fish have none at all, reptiles and birds have but a poor beginning. The cortex is connected to the body by subcortical structures that regulate life-supporting systems, for example, hormonal activity and primal emotions. They connect the cerebral neocortex to the brain stem which in turn connects the brain to the spinal cord and the biological processes of the body. The cerebral neocortex is an important centre of sensory and motor control. It is in the cerebral neocortex that *mind and body* come together and create our self-image and our view of the world.

18 Is seems that Persinger understands God as a personal or non-personal God distinct from the world.

19 Persinger (1987) 72–75.

religion. Furthermore, since, due to this God Experience people may feel reborn, and since to be born again is a soteriological issue in most religions, people who are not religious from the start may join a church or another religious community. However, what actually happens, according to Persinger, is that fatigue and anxiety cause a temporal lobe transient triggering the God Experience.[20] He writes:

> [S]pontaneous intermittent above-normal right hemispheric metabolism could encourage intense brief intrusions that would ultimately kindle or elevate [...] the left hemispheric function. The phenomenological correlates would include enhanced self-esteem, 'restructuring of the self-concept' and sudden understanding. Emergence of 'the new self' following religious conversions is not unusual and often accompanies a peak or oceanic experience that is subsequent to significant psychological depression.[21]

That such a God Experience (spontaneously occurring or not) may give the experiencer a sudden understanding of, for example, the purpose of her or his life is, according to Persinger, due to an intercalation of the hemispheres.[22] What happens is that all fear is gone, the person is no longer depressed and has regained the strength to cope with the situation. According to Persinger there are two periods in a human being's life when the occurrence of God Experiences are more common: (1) During temporal lobe instability around puberty and (2) in the autumn of one's life, due to sclerotic diseases and disorders of blood circulation. As already mentioned, that people have God Experiences is due to the fact that the human temporal lobes are shaped the way they are. Thus, since our brain is constructed the way it is, it is not unusual that human beings undergo God Experiences now and then.

---

20 Since Persinger is not clear on this point, I would like to add that religious experiences do not occur spontaneously only during psychological crises. They also occur in connection with intensive prayer and meditation. I suggest that when a person undergoes a psychological crisis, she or he may change her or his lifestyle, which often implies reflection and/or meditation. For instance, pharmacies supply special music CD's to facilitate reflection and meditation. Sounds such as water falls, waves of water on the beach, etc, are said to have a calming influence on the stressed soul. The person may also be extremely anxious and start to pray or meditate intensively for days on end. The period of praying may have its end in a God Experience.

21 Persinger (1993c) 924. Kindling will be carefully discussed below. Suffice it, for now, to mention that kindling is a clinical way of generating epileptic seizures.

22 Persinger (1993c) 916. Intercalation is defined as the inclusion of neuronal sequences from the other hemisphere within a time frame that would allow both recognition and discrimination. Intercalation is based on the idea that cells have positional information as to where they are supposed to be in reference to neighboring cells. Each cell *knows* where its neighboring cell is supposed to be, for example, cell 1 is supposed to be next to cell 2 which is supposed to be next to cell 3 etc. Imagine that we try to put cell 1 next to cell 6. Both cell 1 and 6 *know* that they are not supposed to be neighbors. Thus a conflict takes place and the cells will start to restore the situation. A mitosis occurs (a kind of cell division) until the positional information of the two cells is correct again. The cells do not care if intercalation results in *normal* morphological patterns, they just want their neighboring cells' positional information to be correct.

3.1.2 Parental Care and God Belief

Why do religious people attach these experiences to a God concept, together forming their God belief? Persinger maintains that "[d]uring the temporal lobe transient that creates the God Experience, body images of early infancy and childhood are transformed into adult formats".[23] Persinger thus maintains that this has to do with the relation that people had with their parents or parent substitutes during their infancy. How?

Neuroscientists maintain that during our first months of life, the temporal lobe stores all information and grows with that information. Later in life, as this temporal lobe shifts to other functions, we forget the stored information. However, Persinger claims that the old information is still there waiting to be remembered as, for example, during a God Experience later in life. He argues that what people experience during such a God Experience is actually their old infant-self. He writes:

> For a moment, the adult concepts of infinity and eternity are experienced with the emotions of the young child. Like the young child, whose universal boundary ends at the edges of his crib, the limits of space and time are blended with the sense of the self. And somewhere outside the twilight boundaries of the cosmic crib, the parental surrogate waits to intervene.[24]

Thus, the adult relives his or her infancy in a modified way, the cradle becomes heaven or the universe, the parent becomes God. Persinger also uses the psychological hypothesis that newborns do not yet have a knowledge of the *I* or the *existential self*.

According to Laura Berk, the self-understanding and self-awareness begins in the second year of life. Before then, the infant does not really experience itself as separated from the surrounding world. However, empirical research has shown that as early as fifteen months, infants separate themselves from their image in the mirror. Some even act silly in front of the mirror, which according to psychologists signifies the beginning of self-recognition.[25] Consequently, if Persinger is right, what the adult relives ought to be his or her infancy before the age of fifteen months.

Basically according to Persinger, God Experiences are reproductions of the dependency on one's parents or parent-substitutes as an infant. From the above we should understand this infancy as ranging from birth to fifteen months.

Persinger argues that children and adults subconsciously attach the qualities and characteristics of their parents to the concept of God. The question

---

23 Persinger (1987) 113.
24 Persinger (1987) 113.
25 According to a study performed by Bullock/Lutkenhaus (1990), Berk (1997) 425.

is how these qualities and characteristics could prepare the infant for a God Experience later in life, especially considering that God Experiences are generated by brain stimulation. Persinger does not answer that question. What he discusses is the child's *behaviour* towards his or her parents or parent-substitutes and the child's and adult's behaviour towards God and people's definition of God, which is a copy of the definition of their parents or parent-substitutes. Thus, if the parents were loving then God is loving. If the parents were conservative then God is conservative. This, however, must imply that the God concept might and probably will change over the years, since people's concept of their parents changes over the years. Furthermore, this should also imply that Persinger must regard the relationship between children who are older than fifteen months and their parents, because the ability to reproduce the characteristics of one's parents to God does not only demand that the infant has a self-recognition but also that it is able to reflect on the self and its environment. Only older children are able to do that.

Persinger calls his psychological mechanism of projecting the characteristics of the parents or parent-substitutes onto God *the principle of association*, and means that how we interpret God is equal to how we interpret our parents or parent-substitutes during our infancy. Let us consider the following example:

Imagine that Veronica is about to drown. She starts swimming as hard as she can and screams *Please God help me!* (loudly or silently). At that moment, following Persinger's reasoning, what she really screams is *mum, dad please help me!* and expects them to do so, because her experience of her parents is such that they always did. Suddenly she sees a log appear which she can hold on to and is finally saved by a passing boat. Veronica believes that God saved her, which of course cannot be the case according to Persinger, because what saved her was a simple log floating in the water. What made her think it was God was the association of her situation to a particular situation in her infancy, replacing her parents with God.[26]

3.2 Neuroscientific Research

Persinger's neuropsychological studies I shall analyse are: Firstly, the one in which he compared the inquiries of persons who practise meditation with persons who do not in order to find a relation between epilepsy and religious experiences;[27] and secondly, a study in which Persinger compares the

26 Persinger (1987) chap. 9.
27 Persinger (1993b).

results of an experiment on religious experiences of right-handed men with those of right-handed women.[28] Along with this, I shall analyse his study of the relation between hemisphere dominance and aggressive religious belief.[29] Finally, I shall address Persinger's neurological study in which he applies the model of vectorial hemisphericity as well as magnetic wave stimulation on the brain in order to find the neural activity that generates religious experiences.[30]

3.2.1 Transcendental Meditation and Kindling

With his study on Transcendental Meditation (TM),[31] Persinger wants to establish that religious experiences are related to *Temporal Lobe Epilepsy, TLE*, because according to him, people who suffer from temporal lobe epilepsy often report having religious experiences. Before presenting his study, I believe it would be of benefit to briefly present Temporal Lobe Epilepsy.

Mark Mumenthaler and Heinrich Mattle define epilepsy as follows:

> Epilepsy is characterized by attacks [seizures] of impaired neurologic function, nearly always combined with loss of consciousness and/or other paroxysmal motor, sensory, or autonomic phenomena.[32] Each seizure is produced by an abnormal electrical excitation of brain tissue that can usually be detected during the seizure as an abnormal pattern on (among others) the electroencephalogram (EEG). There are different types of epilepsy that are caused by different structural and functional (metabolic) anomalies of the brain.[33]

Epilepsy is caused by abnormal or altered neural activity in specific parts of the brain. However, there are several types of epilepsy which are characterized by the type of seizures or by the brain areas where the seizures originate. For our purpose it will suffice to consider one type, temporal lobe epilepsy (TLE). There are two forms of temporal lobe epilepsy, *medial temporal lobe epilepsy* and *lateral temporal lobe epilepsy*. The seizures of the first form originate in the amygdala-hippocampus complex (limbic area) and are called *complex partial seizures* while the latter seizures have their starting point in the lateral temporal cortex. Previously, when neurologists talked about TLE, one meant medial TLE. Since Persinger does not specify

---

28 Persinger (2003).
29 Persinger (1997).
30 Persinger (1993c).
31 Persinger (1993b).
32 Paroxysmal here means through seizures in the motor, sensory or autonomic structures of the brain.
33 Mumenthaler/Mattle (2004) 493.

temporal lobe epilepsy (sometimes he does not specify epilepsy at all), but since he mentions complex partial seizures, he probably intends medial temporal lobe epilepsy (MTLE).[34] The hallmarks of MTLE seizures are that the person, during such a seizure, experiences a "twilight state and clouding (but not total loss) of consciousness".[35] By twilight states neurologists mean that the person performs highly complex activities in an apparently organised manner and remains conscious. This behaviour can go on for days but the person cannot remember anything of this afterwards.

Medial temporal lobe epilepsy may first appear in childhood or in adult life but most commonly appears between the age of 10 and 20. The clinical features are as follows. The seizure has a sudden start and lasts from less than a minute to several hours and sometimes days and then gradually subsides. The person may have "a sensation of warmth or discomfort ascending from the stomach into the neck, nausea or olfactory".[36] The person may have "hallucinations, [experience] a dream-like state, anxiety, [have] a feeling of familiarity déjà-vu or unfamiliarity (jamais-vu)".[37] These experiences are called *auras* or *warnings*. They may be followed by different characteristic behaviours, such as, wandering about, automatisms (for example, lip smashing and rubbing the hands together), grimacing, an arrest reaction (children, for example, will pause in whatever they are doing and will not respond if spoken to), and a fixed, vacant stare.[38]

The other form of temporal lobe epilepsy, lateral temporal lobe epilepsy (LTLE), often causes auditory or visual hallucinations, or dizziness. According to Mumenthaler and Mattle, after the aura(s), the person may feel disorientated, move his or her head to one side, exhibit hand or leg automatisms, and may have aphasia.[39]

Let us return to Persinger's study of the relation between epilepsy and religious experiences. How did he proceed?

Method

Persinger's method was as follows. He collected data from the Personal Philosophy Inventories (PPI) of university students (who did not know anything about the purpose of the study) as the basis for this neurospsychological study. The PPI contained questions to be answered by yes or no.

---

34 Persinger called what is experienced during epileptic seizures complex partial epileptic like signs (CPELS).

35 Mumenthaler/Mattle (2004) 524.

36 Mumenthaler/Mattle (2004) 524.

37 Mumenthaler/Mattle (2004) 524.

38 The clinical manifestations of MTLE is presented in Mumenthaler/Mattle (2004) 525.

39 Mumenthaler/Mattle (2004) 527.

Unfortunately, we are not provided with any examples of the questions of the PPI. However, we know that they were tailored from the reports of a group of epileptic persons concerning what they experienced during epileptic seizures. For example, these persons would say that they had experienced a sensed presence, heard voices, seen a light and so on. Persinger called these characteristics of seizure experiences *complex partial epileptic-like signs* (CPELS). Persinger wants to be able to compare students who are used to meditation with students who are not and to search for similarities between meditation and complex partial epileptic-like signs (CPELS). The result of the PPI of 221 university students, men and women between 18 and 60 years of age who had learned to meditate (most of them, about 65 per cent to 70 per cent Transcendental Meditation (TM), were compared with those of 860 non-meditators. The data was collected between 1981 and 1992. SPSS software was used on a VAX 4000 computer.[40] Moreover, a two-way analysis was performed, namely, gender-meditation, i.e., how women and men respectively respond to meditation.

Persinger's experiment consisted of two parts. Firstly, he had to define the criteria that would allow him to compare religious experiences with epilepsy in order to find out whether or not, and if so, how, religious experiences are related to epilepsy. Therefore, he collected the data from clinical reports on how epileptic persons behave during and after their seizures and of the complex partial epileptic-like signs that were reported by epileptic persons. For instance, the following CPELS were taken into account, (1) experiencing paranormal phenomena, (2) automatic behaviours, (3) writing (keeping notes about personal thoughts), (4) profound experiences from reading or reciting poetry and prose, (5) sensed presence, (6) visual anomalies (visions), (7) auditory anomalies (hearing an inner voice) and (8) sensations of cosmic consciousness.[41]

The second part concerned the significance of the number of years of meditation for the effect of meditation on the brain. Therefore, 56 students out of the 221 meditators, who reported that they practise TM, were compared with non-meditators. Most of the questionnaires concerning the second part were collected between 1981 and 1984. The years of meditation practice were classified into one year, two years, three and a half years and more than five years. The consistency of practice was not analysed.

---

40 SPSS is a software tool for creating statistics and a VAX 4000 is a desktop computer manufactured by Digital Equipment Corporation.

41 The question whether it is not obvious that the experiences 5, 6 and 7, for example, are a common form of religious experience and are hence part of the experiences of religious people, will be considered later.

Hypotheses

Persinger's overall idea is that if we could understand which neural activity generates temporal lobe epileptic seizure experiences, we would know which neural activity causes religious experiences. He seems to assume that there is altered neural activity in the temporal lobes, causing religious experiences. He presents the following two additional hypotheses:

1. *We should expect a larger number of CPELS among those who meditate than among those who do not meditate*, because meditation and especially Transcendental Meditation (TM) are operationally a cognitive kindling.[42]
2. We should expect to find that the occurrence of CPELS increases with the number of years that meditators have been intensively meditating, because repeated meditation is associated with an increase of these signs.[43]

Results and Discussion

Before discussing the study, it is important to know that cognitive kindling refers to a neuroscientific method used in the research on epilepsy.[44] It has to do with provoking epileptic seizures by way of stimulating the amygdala with weak electrical or chemical stimuli. What happens is that after 5 to 10 repetitions, very little provocation is needed to start an epileptic seizure.

The first hypothesis, that a larger number of CPELS was expected among the meditators than among the non-meditators was confirmed. The analysis of the PPI showed that people who meditate reported significantly more complex partial epileptic-like signs than people who do not meditate. For example, they reported having sensed presence experiences, they heard voices, experienced being one with the cosmos, or they had visions. The PPI of the meditating women contained more CPELS than the PPI of the meditating men.[45] There were no significant differences between meditators and non-meditators with respect to olfactory experiences, control, and perseverative experiences. Nothing was said about whether the epileptic-like signs increase in significance or in quantity.

The second hypothesis, that the duration of meditation and meditation techniques, especially TM, encourage CPELS, was also confirmed. TM practitioners clearly showed more epileptic-like signs than non-meditating people. They reported that they frequently experienced vibrations, heard

42 Persinger (1993b) 80.
43 Persinger (1993b) 81.
44 See introduction.
45 See also the study below.

their names called and experienced paranormal phenomena.[46] Furthermore, there was a strong correlation between the number of years of TM practice and the number of the CPELS, i.e., the more years of meditation, the more CPELS were reported. By correlation I mean the joint relationship between two variables, here the number of years of intense meditation and the number of CPELS. The years of TM practice and the number of the epileptic-like signs go together.[47]

Persinger's conclusion and thus the answer to his main question is that religious experiences and epilepsy are related in the sense that (a) people who had religious experiences and people who had epileptic seizures remember their experiences in a similar way. Persinger defines this as complex partial epileptic-like signs (CPELS), (b) both religious experiences and epileptic seizure experiences are caused by cognitive kindling, (c) both religious and epileptic seizure experiences are associated with altered right temporal lobe activity.

Let me try to reconstruct Persinger's thinking:

1. Epileptic persons frequently report having religious experiences called CPELS (hearing voices, having visions).
2. Epileptic seizures are caused by cognitive kindling in the right temporal lobe activity causing above-normal neural activity.
3. The PPI (based on the reports above) showed that persons who meditate have religious experiences (CPELS).
4. *Therefore*, both religious experiences and epileptic seizure experiences are caused by cognitive kindling and are associated with above-normal right temporal lobe activity.

Whether or not the conclusion follows from the premises will be shown in the next chapter. Additionally, as Persinger's study seems to have shown, complex partial epileptic-like signs increase (in number) with the number of years of meditation.[48] Whether Persinger's argumentation is reasonable or not, holds or does not hold, will be shown in the evaluation chapter.

Persinger did not only study persons who suffer from epilepsy but also those who suffer from other types of mental illnesses, for example, schizophrenia. That religious experiences, epileptic seizure experiences and schizophrenic hallucinations are related to each other does not imply that there are no differences between them. Even though Persinger actually suggests

46 Experiences that are beyond the scope of normal scientific understanding.

47 For more information on how correlation is used within psychological research see Liebert/Liebert (1995) chap. 4.

48 Probably they will also increase in intensity, but nothing is mentioned about this.

that most of the world's religious leaders suffer from temporal lobe abnormalities, he maintains that, unlike epileptic persons, the periods of CPELS for healthy religious people are rather brief. Concerning epileptic persons, such signs are constantly triggered. Thus there is a difference in frequency of the experiences. The difference between healthy people and schizophrenic persons who have religious experiences (schizophrenic hallucinations) lies in the functionality of the brain (malfunctioning of the intercalation process in the brains of schizophrenic persons).

Concerning the difference between schizophrenic persons who undergo religious experiences and healthy persons, he writes that

> [t]he primary difference between the dissociative, creative poet, who transiently experiences these intrusions [religious experiences], and the patient with a chronic left temporal lobe focus is [...] [w]hereas the poet can conceptualize about the universe or God and remain intact, the schizophrenic confuses and ultimately loses the boundaries [between subjective experiences and the physical world].[49]

What happens according to Persinger is the following. During the early stages of schizophrenia, the person experiences mild euphoria and enhanced sensory sensitivity. According to Persinger, it is during this period that religious experiences are to be expected. Thus if a schizophrenic person has religious experiences, it will be during this stage of the illness. However, due to the death of certain brain cells, inter-hemispheric intercalation diminishes and the analysing capacity of the left hemisphere is replaced by flat affect and minimal cognition. It is at this stage that the persons get confused. Healthy people's brains do not have decreasing inter-hemispheric intercalation which implies that they are able to analyse and interpret their religious experiences. Thus there is also a difference in brain functionality.[50] Nevertheless, there are of course exceptions. John F. Nash, who received the Nobel award in Economy, 1994, was struck by paranoid schizophrenia in the late Fifties but gradually recovered and could return to his work in mathematics.

Hence, even though Persinger studies these persons, especially their seizure experiences, and he associates their seizure experiences with religious experiences, he does not completely neglect the differences in that the healthy experiencer will be able to benefit from his or her religious experience while the mentally ill person will only be confused. Furthermore, although meditation may enhance CPELS, Persinger emphasises that moderate elevation of these signs may convey creativity and suggestibility, something which might facilitate adaptation.[51]

---

49 Persinger (1993c) 923.

50 Persinger (1993c) 922.

51 That religious people benefit from their religious experiences is an important issue for Persinger's research on religious experience, for better or worse. The next study will show this clearly.

3.2.2 Right-handedness and Religious Experiences

Since more women than men reported that they had religious experiences, Persinger performed another study in which he studied the differences between women and men's religious experiences and focused on the implications religious experiences may have for the self of women and men.[52] It is not clear what Persinger means by *self* but I suggest that he intends the more complex concept of the self of the adult compared to the concept of the self of a child, i.e., their identity. This complex concept of the self or identity has to do with a person's behaviour. According to Baldwin and Baldwin, a person gets his or her identity when he or she learns "self-descriptions that (1) hold true over a broad range of situations and (2) reveal consistent patterns of interests and behaviors".[53] One's identity in psychological terms has to do with self-image, self-esteem and individuation, i.e., the processes whereby the undifferentiated (a new-born) develops individual characteristics. It may also be the process by which components of an individual are integrated into a more indivisible whole (children or persons recovering from having been imbalanced or depressed). However, from the context of the research reports, I believe that Persinger rather intends the behaviour of women and men rather than their self.

Closely related to this study, is the one in which Persinger considers the role of gender and God Experiences in statements such as "I would kill in God's name".[54] Therefore I shall consider both studies together under "Results and Discussion".

Persinger's aim of the study is to see what impact religious experiences have on the behaviour of men and women, whether or not there is a difference in consequences.

Method

Persinger combined the results of a Personal Philosophy Inventory (PPI), similar to the one mentioned above, i.e., based on complex partial epileptic-like signs and the Hemisphericity Questionnaire (HQ), which is a questionnaire for testing hemispheric dominance, of 50 right-handed male and 50 right-handed female students with the results of a neuroscientific experiment performed on these students. All the participants of the experiment were first-year undergraduate psychology students between the ages of 18 and 25 and all were right-handed as defined by their hand preference for writing. There was no left-handed control group, because a previous neuro-

52 Persinger (2003).
53 Baldwin/Baldwin (1998) 350.
54 Persinger (1997).

scientific study suggested that left-handed men are similar to right-handed women and that left-handed women are similar to right-handed men.[55] The students completed a PPI containing yes or no questions, which in previous studies had been associated with temporal lobe activity (i.e., CPELS). The PPI also included questions related to "traditional and exotic religious beliefs".[56] There is, unfortunately, no list of the questions included in the research report but, from what is said in the results, we know that the following questions, for example, must have been included: Do you belong to a certain religious community? How often do you visit your church or religious community? If God told you to, would you kill for God's sake? The HQ contained questions related to which hemisphere is dominant, i.e., questions related to logical thinking and questions related to intuition. The HQ, then, is "a score for a person along a continuum that has been inferred to reflect dominant left hemispheric cognitive styles (logical) to dominant right hemispheric (intuitive) cognitive styles".[57]

After two to six weeks, each student was tested separately. They were asked to sit in a comfortable armchair and were told that the experiment had to do with relaxation. (They were able to leave the experiment at any time.) When they were comfortable, the temporal lobes of the students were stimulated with magnetic waves by way of the Transcranial Magnetic Stimulation apparatus (TMS) which is, roughly, a modified motorcycle helmet, embedded with four sets of solenoids arranged along the temporal plane.[58] It is important to know that the TMS that Persinger applies is not a tool for measurement, observation or diagnostics but a therapeutic and treatment tool for certain psychiatric disorders. Hence it does not measure but actually alters neural activity or values in the brain. For example,

> the effects of TMS on speech are well known and it is now clear that TMS can induce speech disruption [...] There have been demonstrations that applying TMS over the left frontal cortex cannot only disrupt speech production but also impair verbal recall.[59]

In a position statement by the college of psychiatrists of Australia and New Zealand one can read that

> [s]eizures can occur with high intensity, high frequency rTMS [...] Guidelines for safety have been published. [...] Though it is noted that some of the earlier reported seizures occurred despite using parameters which were within the later published guidelines.[60]

---

55 Gordon (1991) referred to by Persinger (2003) 13.
56 Persinger (2003) 8.
57 Persinger (2003) 8.
58 Solenoids are coils of wire that can be used to generate electromagnetic fields.
59 Walsh/Cowey (2000) 77.
60 The Royal Australian and New Zealand College of Psychiatry (2003) point 7.

One might perhaps say that TMS is a kind of kindling. Needless to say this raises questions concerning what kind of experiences the participants of Persinger's experiment actually had. According to the writers of these reports, the application of the TMS within psychiatry is still discussed and questioned.

The experiment was performed as follows. The participants were asked to sit in a chair. Their eyes were covered with half of a ping-pong ball containing cotton. The modified helmet was placed around their heads. A specific amygdaloid burst-firing like pattern was generated by a computer. This magnetic pattern was presented to regions of the temporoparietal cortices of the persons as 5 minutes off – 10 minutes on intervals during 35 minutes. Persinger selected this longer periodicity, because cells often habituate to continuous magnetic field presentation within about 15 minutes and because he considered repeated on-off episodes more optimal to encourage the types of transients that are associated with the experience of sensed presence. After 35 minutes the students were asked to complete a questionnaire that began with the following instructions.[61]

> While you were sitting and were wearing goggles, some stimuli were presented. To determine whether you detected them, please indicate the incidence (frequency) of the experience that occurred from the beginning of this experiment by circling the appropriate numbers: 0 indicates *never*, 1 indicates *occasionally* (or at least once], and 2 indicates *frequently*.[62]

Hypotheses

Persinger put forward the following hypotheses.

1. We should expect that *right-handed women have more frequently sensed presence experiences than right-handed men*, because (a) in a previous study women reported more often that they had such experiences than men did and (b) another study showed that such experiences are mostly associated with right-hemispheric dominance. Right-handed women and left-handed men are associated with right-hemispheric dominance.[63]
2. We should expect that right-handed men who have sensed presence experiences will interpret them as something foreign to the self, whereas right-handed women interpret these experiences as belonging to the self or as feeling united, right-handed men may interpret their sensed presence experience as an experience of another mind or

61 Persinger (2003) 8–10.

62 Persinger (2003) 9.

63 Persinger (2003) 7.

> of a spirit. Right-handed women, on the other hand may interpret such a sensed presence experience as being absorbed in a wonderful light. This is due to their left-hemispheric dominance and right-hemispheric dominance respectively.[64]

An implicit hypothesis that Persinger also uses is, that from an evolutionary point of view, left hemispheric men i.e., right-handed men, should experience less fear and should have an elevated belief in their own survival than should right hemispheric men i.e., left-handed men. Similarly, left-hemispheric women, i.e., left-handed women should experience less fear and should have an elevated belief in their own survival than do right hemispheric women, i.e., right-handed women. According to evolutionary explanations, these capacities are meant to facilitate the defence of the group during, for example, warfare. Right hemispheric women i.e., right-handed women, (and similarly right hemispheric men, i.e., left-handed men) on the other hand, are vigilant and intuitive, which facilitates the protection and preservation of the children. Thus right-handed men (and similarly left-handed women) *defend* for the sake of survival while right-handed women (and similarly left-handed men) *protect* for the sake of survival.

Results and Discussion

It was shown that right-handed women have more frequent experiences of relative intrusion than right-handed men (a mean of 0,21 for women and 0,16 for men), of relative presence (a mean of 0,54 for women and 0, 25 for men), of fear or terror (a mean of 0,56 for women and 0,42 for men), of feeling dizzy or odd (a mean of 0,62 for women and 0,10 for men), of vibrations (a mean of 0,22 for women and -0,2 for men), of out-of-body-like-experiences (a mean of 0,68 for women and 0,8 for men), of feeling somewhere else (a mean of 0,74 for women and 0, 45 for men), of spinning around (a mean of 0, 68 for women and -0,1 for men) and of dream images (a mean of 0,31 for women and -0,10 for men). Right-handed men, more than right-handed women, have frequent experiences of a sensed presence (a mean of 0,94 for men and 0,65 for women), of ego-alien intrusions (0,83 for men and 0,06 for women). Persinger therefore considered both hypotheses confirmed, namely, that right-handed women more frequently than right-handed women have sensed presence experiences and that right-handed men who have sensed presence experiences interpret them as something foreign to the self.

Persinger maintained that the latter, combined with the implicit evolutionary assumption, has the following consequence: Right-handed men (and

---

64 Persinger (2003) 7.

left-handed women) who have such a sensed presence experience, may use this experience as "justification for aggression and for the mass slaughter of other human beings because of divine consent".[65] According to him "[t]he experience of 'divine consent' is often described as an overpowering conviction that an idea has been condoned by a sensed presence. It is attributed by the experiencer to an omnipotent extracorporeal being that is perceived as boundless in time [eternal] and space [omnipresent]".[66] Persinger justified his conclusion by referring to the compilation of the PPI of his study which shows that seven to eight percent of the male students admit that they would kill other people if "Allah or Jehovah" told them to kill.[67] This percentage increased to 44 percent when the respondents were right-handed men (but not right-handed women) who attend church regularly. These data, for example, *God told me to kill these persons, or God told me to go to war with a certain group of people*, underlie Persinger's conclusion that right-handed men, who are left-hemispheric dominant, are potentially dangerous persons if they should have a religious experience. This is not the case for right-handed women.[68]

We could reconstruct Persinger's arguement as follows.

1. Right-handed men experience less fear, have a strong belief in their survival which facilitates the defence of the group during warfare.
2. Right-handed men and left-handed women are considered equal concerning hemispheric dominance.
3. 7% to 8% of the not churchgoing right-handed men in the study answered yes to the question whether they would kill other people if God told them to do so.
4. 44% of the churchgoing right-handed men in the study answered yes to the question whether they would kill other people if God told them to do so.
5. *Therefore*, right-handed men (and thus left-handed women) are potentially dangerous persons if they should have such a religious experience.

In another study entitled "'I Would Kill in God's Name': Role of Sex, Weekly Church Attendance, Report of a Religious Experience, and Limbic Lability",[69] Persinger writes:

---

65 Persinger (2003) 14.
66 Persinger (2003) 14. Since Persinger refers to the sensed presence as an omnipotent extracorporeal being, we have an example of Persinger's understanding God as distinct from the world.
67 Persinger (2003) 14.
68 Persinger (1997)/(2003).
69 Persinger (1997).

> Wars, a form of group conflict within *homo sapiens*, have been responsible for the death of more than 100 million people within the last century. Although wars may be driven by economic and political variables, the capacity to maintain the group dynamics is strongly determined by the manipulation of religious beliefs.[70]

It seems that Persinger blames religious beliefs as the main cause of war and the death of millions of people. In the first study accounted for above, he concluded that religious experiences are caused by cognitive kindling. This might be the reason behind his idea that people who share the same religious belief can be stimulated towards political ends. He claims that a crowd's mood can easily be manipulated by means of, for example, music. Music has a stimulating effect upon the temporal lobes, which Persinger argues, may trigger religious experiences. Thus, following the second study above, if we assume that music triggers religious experiences in right-handed men, then this experience could be remembered as *feeling the hand of God*. In such a way, Persinger maintains, people can be brought to kill other people in the name of God. His arguement goes as follows:

1. Religious experiences are caused by cognitive kindling which is a kind of brain stimulation.
2. Music is a good instrument for stimulation.
3. Music stimulates the temporal lobes which may trigger religious experiences.
4. A group of people who share the same religious belief can be stimulated towards political ends such as war.
5. Right-handed men (and left-handed women) interpret their religious experiences as coming from something alien to the self, a spirit or another mind.
6. The religious experience may be remembered as being moved by the spirit.
7. *Therefore*, a group of people who share the same religious belief may kill other people in the name of God.

Needless to say, these studies and the conclusions that Persinger draws are controversial. Therefore profound attention will be given to them in the evaluation chapter.

70 Persinger (1997) 128.

3.2.3 Vectorial Hemisphericity

The last study to be analysed is the one in which Persinger applies the model of vectorial hemisphericity as well as magnetic wave stimulations on the brain.[71] He now wanted to study the significance of the hemispheres for religious experience. He maintains that the

> isolation and understanding of the neuro-electrical bases of mystical experiences and religious conversions are important in the light of their powerful modifications of the sense of the self and the subsequent change in beliefs and behavior. If their bases were understood and hence could be generated experimentally, application to clinical intervention for the treatment of depression, anxiety, and dissociation could become a powerful tool.[72]

Persinger's aim is hence to find the neural activity that generates religious experiences, because this information may be important for clinical use. Persinger maintains that "[m]ultiple variants of sensed presence often precede mystical and religious experiences that are frequently followed by sudden, permanent changes in the self-concept".[73] According to him, this could be why religious persons seem to be less depressed than non-religious persons and why, if they are, religious persons seem to recover more quickly. Persinger suggests that the processes that generate religious experiences could be copied and artificially transmitted to persons suffering from clinical depression in order to alleviate the consequences of the illness.[74]

Method

The model of Vectorial Cerebral Hemisphericity is used because Persinger believes that the hemisphere in which the above-normal neural activity occurs (right or left hemisphere), will determine the affect, content and type of experience. Furthermore, depending upon the above-normal neural activity of the two hemispheres (the intercalation), intrusions of the right hemisphere (sensed presences) and equivalent of the left hemisphere (linguistics), the person will have an experience of "evil entities, gods, out-of-body experiences, and alterations in space-time".[75] The Vectorial Cerebral Hemisphericity explanatory model identifies neural processes and is, roughly, a kind of map of the brain and its hemispheric structures and functions.[76] The

---

71 Persinger (1993c).

72 Persinger (1993c) 926–927.

73 Persinger (1993c) 915.

74 Persinger uses the term anxious depression instead of clinical depression.

75 Persinger (1993c) 915.

76 This *map* was drawn while studying persons whose corpus collosum was cut. Neuroscientists discovered that different neurofunctions are localized in different hemispheres. However, there is

explanatory model of vectorial hemisphericity is a map of the brain that illustrates three states of neural activity of the left and right hemispheres separately: below normal, normal and above normal.[77] The model is used along with PET-imaging technology and the Transcranial Magnetic Stimulation apparatus (TMS), *the Helmet*. As mentioned earlier, Persinger does not use PET-imaging technology to study persons undergoing religious experiences. Rather, he uses clinical reports of the results of such PET-imaging of persons who had epileptic seizure experiences and persons who experienced schizophrenic hallucinations. He studies hemispheric neural activity by way of kindling the temporal lobes of the participants with the TMS.

This neuroscientific report does not mention anything about the participants. From the description of the study we can guess that the participants were investigated individually. How many they were, their ages, gender etc. is not known.

Hypotheses

Persinger proposes the following hypotheses.

1. We should expect that stimulation of the right hemisphere with TMS magnetic waves will provoke sensed presences and fear-like experiences, because the right hemisphere subserves vigilance and its more extreme variant (anxiety, fear, or panic), those experiences should be

---

no clear distinction between the neural activity in the left and right hemisphere. Everything we do or experience is associated with complex neural activity in the entire brain area. Nonetheless, neuroscientists agree that the right hemisphere is more associated with non-verbal thought, visual spatial perceptions, and the perception, modulation and expression of emotions than the left hemisphere, and that the left cerebral hemisphere is more analytically inclined, and is more associated with language and mathematical processes than is the right hemisphere. Religious experiences are associated with the right hemisphere for what concerns, for example, visions, hearing a voice or feelings, and with the left hemisphere for what concerns the interpretations (language and logical and rational thinking). When the corpus collosum of epileptic persons is cut to prevent them from having epileptic seizures, their brain becomes what neuroscientists call a *split brain*. These persons view their environment in the following way. If a split-brain-person is shown some object, say a pencil, in such a way that the visual input can only get in touch with the left side of the brain, he or she will tell you exactly what object he or she sees, i.e., a pencil. If the pencil is shown to the person in such a way that the visual input can only get in touch with the right side of the brain, he or she will be able to draw what he or she has seen but will not be able to add the right words, he or she will not be able to say that the drawing represents a pencil. The reason for this is that the ability of the left side to turn visual input into logically verbalized concepts is lacking, i.e., there is no intercalation. It is from such research results that the conclusion is drawn that each hemisphere is capable of some kind of awareness, but that they experience and express it in very different ways.

77 What is meant by below normal, normal and above normal, is purely neuropsychological. *Normal* simply denotes what is neuropsychologically understood as normal activity of this or that part of the brain.

primarily negative and should be enhanced within partially sensory deprived environments.[78]

2. We should expect that stimulation of the left hemisphere with TMS magnetic waves should provoke elevated self-esteem and no fear, because above normal neural activity in the left hemisphere is not associated with religious experiences or fear, because the left hemisphere is said to be the seat of human linguistic and rational capacities and no emotions should be involved.[79]
3. We should expect that "direct stimulation of the middle thalamic nuclei by amygdaloid input through the inferior thalamic peduncle could simultaneously but differently generate signals to both hemispheres" and encourage inter-hemispheric intercalation. We should then expect experiences of entities, intrusive voices, etc, and fear to be reported. This is due to the involvement of the right hemisphere.[80]

Results and Discussion

For the sake of the clarity of the analysis, I shall divide religious experiences into three different types. Since the English language does not have different words to distinguish different kinds of religious experience, I shall use the German words *Erlebnis* and *Erfahrung* for my purpose.[81] I believe that this distinction will prove fruitful for how we should understand Persinger's results and the discussion of his study.[82] The three kinds of religious experience I have in mind are: Religious Erlebnis (RErl), Religious Erfahrung (RErf) and Religious Ideological Experience (RIE). RErl is the Religious experience *per se*, the very moment of experiencing before the experiencer put the experience into words. It is those experiences that Persinger associated with right-hemispheric above-normal neural activity.[83] RErf is a RErl extended with an interpretation. RErf develop over the years and may become very broad and complex. Persinger associated these experiences with hemispheric intercalation. A RIE does not need to include any RErf at all. RIE are based on logical and rational reflections. Those

78 Persinger (1993c) 919.

79 Persinger (1993c) 918.

80 Persinger (1993c) 918.

81 The Dutch words *belevenis* and *ervaring* or the Swedish words *upplevelse* and *erfarenhet* would also do but might be less known internationally.

82 The division made is quite different from more established ones: for example, John Smith in Experience, Analogy and Religious Insight or Wayne Proudfoot in Religion and Inquiry in William James.

83 I am aware of the fact that as soon as one talks about an experience it is conceptualized and loses its RErl status. However, I still think that it is fruitful to make the distinction. Doing so I am able to clarify Persinger's and Newberg and d'Aquili's study.

experiences are related to left-hemispheric above-normal neural activity. Obviously, people may belong to a certain religion merely because it has important advantages. Thus religious Erfahrung (RErf) include religious Erlebnis (RErl) while religious ideological experiences (RIE) may include RErl but the RErl are not essential. RErf includes both RIE and RErl.

The results were as follows. According to some of the participants of the experiment, when their right temporal lobe was stimulated by way of the TMS, they often experienced fear in combination with a sensed presence experience. Such right hemispheric intrusions are then described by the experiencer as impressions, entities, cosmic consciousness, spiritual beings, etc. (RErl). Persinger concluded that these experiences, reflecting the major function of the right hemisphere, are maximally affective and minimally linguistic, and more often interpreted as unpleasant and fearful than as pleasant and secure. Thus the first hypothesis seems to be confirmed. This was not the case when their left temporal lobe was stimulated. These thoroughly left hemispheric experiences are maximally linguistic and minimally affective and are not fearful (RIE). Hence, the second hypothesis was also confirmed. The third hypothesis, that the person would experience sensed presences with experiences of fear was not confirmed. Rather, the opposite occurred. No fear at all was experienced (RErf).

Persinger does not say why the third hypothesis was not confirmed. However, Carl G. Jung might shed some light on the matter by maintaining that some religious experiences (RErl) can be so powerful that they can destroy the experiencer totally (with anxiety) if the experiencer does not have the adequate religious frame of reference to encapsulate it (given by left hemispheric functions).[84] When such a frame of reference or possibility for interpretation is not available, the experiencer may become completely mentally deranged. Persinger might thus have referred to the analytical capacity of the left-hemisphere as a reason why the third hypothesis was not confirmed, because this could be the reason why the experiencer is able to rationally understand his or her religious experience which causes his or her fear to disappear. These experiences hence are linked to both the part of the brain (right hemisphere) related to religious Erlebnis (RErl) as well as to the part of the brain that is related to experiences and beliefs obtained through traditional and logical thinking (RIE.) Together the RErl and the RIE form a rational religious Erfahrung (RErf).[85] Persinger suggests that this intercala-

84 Jung (1969) 316ff/Costello (1989).

85 Almost every human activity that includes images and pictures which are in need of logic and/or language to be communicated, needs the cooperation of both hemispheres. This is also true for religious experiences. For instance, to enable mystics to report about their mystical experiences, the left side of their brain will analyse the vision and will search for a logical and rational explanation, while the right side of their brain will regard the vision as a whole.

tion might even be the reason why some persons suddenly and unexpectedly regain their health. He writes:

> The differential activation of the left and right hemispheric processes could also explain the immunological changes that contribute to altered health following some mystical and religious experiences. [...] such immunofacilitation could also accommodate the health improvements that are reported to follow some religious conversions and that accompany positive changes in self-concept, self-esteem, and other correlates of enhanced left hemispheric metabolism.[86]

Thus, as suggested earlier, Persinger also believes that religious beliefs have positive consequences for human beings, because these may help people, if not to overcome their illnesses, at least to improve their self-concept and attitude towards life. Persinger's suggestion could be reconstructed as follows. Since pleasant mood is associated with a normally activated left hemisphere and anxiety with an above-normally activated right hemisphere and furthermore, since above-normal neural activity in the left hemisphere should be associated with high self-esteem,[87] these assumptions could be of benefit for psychiatric study of clinical depression. From such a perspective, any process that enhances left-hemispheric activity might reduce depression. According to Persinger, talk therapies would be such a process. The most effective method would be, according to him, if it were possible to artificially increase the neural activity in the left hemisphere by, for example, TMS. However, stimulation of the right hemisphere would also be of benefit according to Persinger. He writes that

> spontaneously intermittent above-normal right hemispheric metabolism [neural activity] could encourage intense brief intrusions that would ultimately kindle or elevate [in a manner analogous to the 10 to 15 trials of electroconvulsive shock] the left hemispheric function. The phenomenological correlates would include enhanced self-esteem, "restructuring of the self-concept" and sudden understanding. Emergence of "the new self" following religious conversions is not unusual and often accompanies a peak or oceanic experience that is subsequent to significant psychological depression.[88]

Thus, (spontaneous) religious or mystical experiences may have a healing effect upon (religious) people because they may get sudden insight into, among other possibilities, the purpose of their lives.[89] Furthermore the combination of a religious or mystical experience and receiving an understanding is neurologically associated with an intercalation of both hemispheres. Persinger concludes that persons who, for example, suffer from right tem-

86 Persinger (1993c) 924–925.
87 Since below-normal left hemispheric neural activity is associated with low self-esteem.
88 Persinger (1993c) 924.
89 The experience need not be spontaneous nor does the person have to be religious.

poral seizures like epileptic persons, could be relieved from the negative effects if their left temporal lobes were stimulated in a way that allows an intercalation which would prevent the negative effects (of anxiety) from appearing. This also implies that some epileptic persons experience anxiety during their seizures, because they have, at the time of the seizure, diminished functionality of the left hemisphere, which healthy people who have a religious experience do not have.[90]

What then are the neural functions and structures that underlie religious experiences? Persinger does not give any detailed answer to this question. He merely states that right hemispheric structures and functions underlie experiences such as sensed presences as well as the experience of anxiety (RErl) while left hemispheric structures and functions do not (RIE). Furthermore Persinger maintains that when an interhemispheric intercalation occurs (RErf), the left hemispheric functions somehow eliminate the anxiety involved in right hemispheric experiences. Hence the brain functions and structures that generate religious experiences are the left and right hemispheres and the neural activity behind the intercalation of the two hemispheres. In other words, almost the whole brain.[91]

3.3 Persinger's Conclusions

What conclusion does Persinger draw from (his) neuropsychological research on religious experience?[92] As should have become clear by now, Persinger's implicit aim for his studies of religious experience is to be able to show that they are illusory and that there is no God or Ultimate Reality. It is his view that all religious belief should be abandoned. For instance, Persinger maintains that

---

90 Persinger associates left-hemispheric dominance with egocentrism and right-hemispheric dominance with allocentrism. He also maintains that left-hemispheric dominant persons (egocentric) are those who, more often than right-hemispheric dominant persons, maintain that "If God told me to kill, I would do it in his name," Persinger (1993c) 921–922. However, because Persinger develops this thought in another study which I have analysed above, I shall not discuss it further. For the same reasons, I shall not go further into his suggestions of the implications of religious experience on right and left-handed persons, Persinger (1993c) 925, nor on the relation between epileptic seizure and religious experiences, Persinger (1993c) 922–923.

91 Let me draw your attention once more to the fact that Persinger argues that the brain *generates* or *triggers* and not *correlates* with such and such experiences. I suggest that we could agree that the functions and structures of the human brain in some complex way correlate with our experiences, actions and reactions. However, this does not mean that it generates our experiences, actions and reactions. I will return to this in the next chapter.

92 With his neurospsychological research I do not intend only the selection I have presented above.

> [t]he process that precipitates God Experiences may also contain some fundamental flaw that could eradicate us from the face of this earth. This potential may have been in our genetic pool for millennia, waiting to be released by a special sequence of events. They would be the sign stimulus to an innate releasing mechanism. It would not be the first time that a behavior on which a species capitalized to survive was also the basis for its destruction.[93]

Thus, the very item (i.e., the God Experience) of survival for the human species could, in another environment, be the very item of its destruction. To illustrate this Persinger depicts what a nuclear-war-situation would possibly look like. Persinger creates the following scenario:

Assume that the world is governed by a few political leaders who control the destiny of humanity. Assume that one of these leaders believes in God and eternal life and that this leader undergoes a God Experience which results in a sudden *knowing*. "He feels that to save humanity from a perceived evil [non-believing political leader and his or her politics], all humanity must die."[94] He relates this sudden conviction to his God Experience and thinks *God told me so*. Consequently, and with the confidence of the inspired knowledge, "he pushes the button [...] and [t]he nightmare is finished".[95]

Persinger thus seems to believe that the end of our world will be caused by a religious believer having a God Experience. He concludes that religious beliefs found within the "objective context of military decisions and political confrontations",[96] should be seriously considered. If I understand Persinger correctly, he actually means that religious beliefs (persons) should not be allowed in the military or in politics.

If we relate this to Persinger's study concerning the implications of religious experiences on the behaviour of men and women mentioned above, it seems that this leader should be a right-handed man or left-handed woman who has had a religious experience which he or she comprehended as *God told them to kill*, because, according to Persinger, these persons interpret their religious experiences as caused by something alien, in this case, God.

However, in *Neuropsychological Bases of God Beliefs*,[97] Persinger seems to regard *all* religious people as potentially dangerous when they have a God or Ultimacy experience (and not merely a specific religious experience) and are in some leading position. We could reconstruct his reasoning as follows:

---

93 Persinger (1987) 136.
94 Persinger (1987) 139.
95 Persinger (1987) 139.
96 Persinger (1987) 140.
97 Persinger (1987)

1. The political leader believes in God and eternal life.
2. He or she has a certain God Experience *telling* him or her that all people must die to save the world.
3. He or she destroys all people.
4. *Therefore*, God-beliefs deserve special consideration. The neural processes that underlie God Experiences must be mapped.

Persinger considers religious belief to be such a potential danger that it has to be studied in an objective context which, according to him, can only be a scientific context. His arguement is that even though he recognises the matter of privacy and integrity, "if a man in a crowd holds a bomb in his personal possession, do we ignore it for the sake of his 'privacy'?"[98] Hence, he means that religious beliefs could be *ticking bombs* and in order to save the human race we have to accept ourselves the way we are, namely as biochemical systems that display behavioural patterns and are subject to the laws of conditioning. Our religious experiences are nothing but products of our brain.

However, there is one problem according to him, namely, that believers are not willing to allow their beliefs to be scientifically investigated, because this would imply that they *will have to* accept that there is no God or Ultimate Reality and that religious experiences are illusory.[99] It is not clear what makes Persinger so certain that they will have to accept that religious experiences are illusory. For example, is it because (a) science can empirically prove that religious experiences are illusory, (b) science can empirically prove that there is no Ultimate Reality or God and therefore religious (God) experiences are not possible, (c) science can neither prove (a) nor (b) but the arguements that such is the case are so strong that one has to accept their validity and thereby reject one's religious beliefs.

According to Persinger, the reason why people believe in God, as already discussed earlier, is because through centuries and in all cultures, death-anxiety has been and still is, reduced by way of God concepts. However, he says that these concepts are nothing but words, and since human beings are governed by language in all aspects, he argues that they should realise the limitation of language. They should realise that some phenomena are nothing more than creations of human language. He maintains that there is nothing that these language creations refer to outside the language itself, which he continues, is especially true for religious phenomena. As a matter of fact, he says, "God" is simply "dog" spelled backwards.[100]

---

98 Persinger (1987) 143.
99 Persinger (1987) 143–144.
100 Persinger (1987) 84.

Persinger continues by claiming that the reason why most people do not realise this is, because they usually do not distinguish the phenomena from its concept since, (1) they do not recognise the distinction between concrete and abstract words, and (2) they do not always realise that not all questions have answers. Furthermore, there is the inability to prove that something does not exist, which is a bad arguement according to him, because we cannot prove anything that does not exist.[101]

For Persinger, words should refer to things that are scientifically measurable. Consequently, words which aim at non-empirical referents such as God, eternity, infinity and love are problematic, because these words do not refer to anything that is measurable by scientific means. Such referents do not exist, except as words, and they can never exist as anything more than words. Persinger maintains that only physical things can be things that are not yet detected. Should they not be physical, they simply could never be detected.

As already mentioned, Persinger's research and the conclusions that he draws are highly controversial. His studies of religious experience are of real interest for evaluation.

3.4 Summary and Conclusions

It seems that the aim of Persinger's neuroscientific research is to determine which portions of the brain generate religious experiences and to illustrate that such experiences are illusory and that there is no God or Ultimate Reality. He maintains that all religious beliefs should be abandoned and claims that most wars and group degradations are coupled implicitly to God beliefs. God is only a word, a creation of language and nothing more.

The conclusion drawn from his neuroscientific studies considered above is that God Experiences occur during personal crises, the most severe being those when a person is confronted with his or her own death or the death of a loved-one. God Experiences were necessary for the survival of the human species, because human beings were constantly exposed to danger and hence, something had to balance these threats. The God Experience is triggered by the person's fatigue, anxiety, meditation or intensive prayer and is neurologically associated with above-normal right hemispheric activity. When an intercalation between the hemispheres occurs, the person may suddenly understand the purpose of her or his life. Fear is gone and she or he, if not yet religious, may suddenly believe in God.

101 Persinger (1987) chap. 7.

However, we know that according to Persinger, believing that God exists is illusory. What happens is that the person relives his or her infancy in a modified way, the cradle becomes heaven or the universe, the parent becomes God. God becomes a substitute for one's parents or parent-substitutes. It was maintained that children and adults subconsciously attach the qualities and characteristics of their parents, for example, loving, caring, and intervening, but also punishing, to the concept of God. This psychological mechanism is known as *the principle of association.*

The conclusions drawn from the neuroscientific research are as follows. Firstly, we saw that he uses the same method for different studies, namely, he compared the answers of the inquiries (Personal Philosophy Inventories – PPI) of persons who practised meditation (mostly TM) with persons who did not. The questions of the PPI were tailored from the reports of a group of epileptic persons on what they experienced during their epileptic seizures. These experiences are called complex partial epileptic-like signs (CPELS). In the latter studies he added the Hemisphericity Questionnaire (HQ) which is a questionnaire for testing hemispheric dominance. He compared the answers derived from the PPI and the HQ with the results of an experiment he performed on the participants, namely, stimulating their temporal lobes with magnetic waves by way of the Transcranial Magnetic Stimulation apparatus (TMS). It was also said that the TMS is not a tool for measurement, observation or diagnostics, but a therapeutic treatment tool for certain psychiatric disorders. The TMS is also used along with the Vectorial hemisphericity model, which is a kind of map of the brain and hemispheric structures and functions. Persinger does not use any neuroimaging technology but takes into consideration the scans taken of the brains of epileptic persons.

Religious experiences and epilepsy are related in the sense that (a) people who had religious experiences and epileptic persons who had epileptic seizures remember their experiences in a similar way, which Persinger defined as complex partial epileptic-like signs (CPELS), (b) both religious and epileptic seizure experiences are caused by cognitive kindling and (c) both religious and epileptic seizure experiences are associated with above-normal right temporal lobe activity. Persinger's idea is that if we could understand which neural activity causes epileptic seizure experiences we would know which neural activity generates religious ones. The difference between healthy people and epileptic persons who have religious experiences (CPELS) lies in the frequency of the experiences. The difference between healthy people and schizophrenic persons who have religious experiences (schizophrenic hallucinations) lies in the functionality of the brain.

Furthermore, Persinger considers right-handed men and left-handed women to be potentially dangerous because they may use religious experiences, which they may interpret as hearing the voice of God, as justification for their aggression. According to Persinger, right-handed women and left-handed men (right hemispheric dominants) have more frequent experiences of extra-normal or religious experiences than right-handed men and left-handed women (left hemispheric dominants). However, the latter group will interpret their experience as an experience of something alien, which is not the case for the former group. Since religious experiences are caused by cognitive kindling of the temporal lobes, it is possible to manipulate people for political interests. It was maintained that this is especially true for right-handed men and left-handed women, since they take their religious experiences to be caused by something alien, and they may interpret their experiences as *God told me to kill.*

Persinger also concluded that if the neural activity that generated religious experiences (RErf) were understood and hence could be used experimentally, it could be applied in the psychiatric treatment of depression, anxiety and dissociation. The result of Persinger's experiment is that if the right hemisphere is stimulated the person has an experience of a sensed presence and experiences anxiety, or a religious Erlebnis (RErl). When the left hemisphere is stimulated, the person does not experience any fear but has a religious ideological experience (RIE). When an intercalation occurs in both hemispheres, the person will still experience a sensed presence but the fear will have disappeared, the RErl has been supplemented with an interpretation and has become a religious Erfahrung (RErf).

Persinger seems to believe that the world will be destroyed by a religious leader. Hence, according to him, something has to be done. People have to understand that there is no God or Ultimate Reality. It is all a matter of language. People have to learn to distinguish between abstract and concrete words. The former are nothing but words, such as those used in religious or love matters. The latter are scientifically measurable and only these words have real value, because Persinger maintains that only these words are anchored in reality.

4. Evaluating Persinger's Research

Having analysed a selection of the research undertaken by Michael Persinger on religious experience, I shall continue to evaluate these studies.[1]

To begin with, it is important to investigate how his neuroscientific research on religious experience is performed from a scientific point of view. Hence, I shall explore whether or not and if so, to what degree, these studies measure what they purport to measure or what Persinger believes his research measures.[2] For instance, does Persinger draw conclusions which concern other issues than those he has examined? Such would be the case if the sample he used for his investigations did not correspond with the population concerned.[3] Such would also be the case if the methods he used were inaccurate or incomplete.

It will also be evaluated whether it is plausible to draw the conclusions Persinger draws. Are epileptic seizure experiences and religious experiences related to each other in the way he claims they are? Are religious people dangerous? Is Persinger's study of Vectorial hemisphericity of relevance for religious experiences? Are God Experiences projections of a person's experiences of her or his parents? Finally are all religious experiences illusory, and is God or Ultimate Reality only a matter of words?

It is important to keep in mind what Persinger maintains in *Neuropsychological Bases of God Beliefs* along with his empirical research, because these studies are interwoven. If they are not always explicitly interwoven, they are implicitly so.[4]

4.1 The Scientific Validity of the Studies

We shall now begin the evaluation of the scientific validity of Persinger's neuropsychological studies of religious experience. Scientific validity means

---

1 Persinger (1993b)/(2003)/(1997)/(1993c).

2 It may be the case that a specific method actually measures what it is meant to measure (for example, sensed presences), however that is not what the researcher thinks it measures (for example, religious experiences).

3 A sample is a representative subset of the population to study. A sample has to be representative of that population, because it is used to test hypotheses about that population. Otherwise, the test would be irrelevant. A population is a segment of the world and is the total collection of people, things or events of interest for a specific study.

4 Persinger (1987).

that the question is asked to what degree empirical studies measure what they purport to measure (or what the scientists believe their studies measure). Such questions are, for example, whether the right type of apparatus or method has been used, whether the information has been accounted for in the correct way, whether the samples of the study were relevant and/or sufficient, whether the population of study was adequate, and whether the research results really indicate what they claim.

4.1.1 Religious Experiences and Epilepsy

As mentioned in the analysis, in Persinger (1993b), Persinger wants to find the relation between religious experiences and epileptic seizure experiences. We saw that his conclusion was that religious experiences and epilepsy are related in the sense that (a) people who have had religious experiences and people who have had epileptic seizures remember their experiences in a similar way (these experiences Persinger defined as *complex partial epileptic-like signs* (CPELS)), (b) both religious experiences and epileptic seizure experiences are caused by cognitive kindling, (c) both religious experiences and epileptic seizures are associated with above-normal right temporal lobe activity.

To reach this conclusion, Persinger compares people who are used to meditating (mostly Transcendental Meditation – TM) with people who are not. The data he used for his comparison came from a questionnaire called the Personal Philosophy Inventories (PPI) that the participants had to complete.

However, the problem is that Persinger's questionnaire is tailored from answers given by epileptic persons who reported on what they experienced during their seizures. The epileptic persons would say that they experienced a sensed presence, heard voices, saw a light, etc. These answers were then transformed into questions that the participants (the meditators and the non-meditators) had to answer with yes or no. For example, such questions could be: Have you ever experienced a sensed presence? Have you ever heard a voice that was not your own, though nobody was speaking to you?

I wonder whether there are not good reasons to suggest that his study is biased. For instance, does not meditation aim at religious experiences and is it not a characteristic of religious experience to feel sensed presences, hear voices or have visions? Hence, most (if not all) of the persons who practise meditation on a regular and intensive basis will probably answer yes to at least one of the questions of the PPI, while persons who do not meditate will mostly answer no to these questions. Hence, when Persinger compares the PPI of the meditating persons with the PPI of the control group of non-meditating persons he can conclude that meditators have significantly more CPELS. For example, they feel a sensed presence or hear a voice more often than the non-meditators. Furthermore, when he compares the PPI of the meditators with the reports of the epileptic persons (on which the questions of the PPI were based) he can conclude that there is no doubt that religious experiences and epileptic seizure experiences are related. His question then is *how* these experiences are related. Since epileptic seizure experiences are caused by cognitive kindling of the temporal lobes, and since both the meditators and the epileptic persons have similar experiences (CPELS), Persinger concludes that religious experiences are also caused by cognitive kindling. Hence, both religious experiences and epileptic seizure experiences are associated with above-normal right temporal lobe activity. Religious experiences and epilepsy are related experiences.

However, should not the questions of the PPI have been selected from testimonies of people belonging to different religious traditions who had religious experiences rather than from people's descriptions of what they experienced during epileptic seizures? In my opinion, Persinger would then have been able to compare descriptions of religious experiences with descriptions of epileptic seizure experiences and investigate whether or not these two kinds of experiences are related to each other. We do not know what other questions the PPI included other than those originating from the reports of the epileptic persons concerning what they experienced during their seizures, and which might make the study as a whole valid. Nevertheless, in my view, the question whether Persinger's studies are scientifically valid is worth raising for the reasons given above.

To see whether my suspicions of the lack of scientific validity of Persinger's neuropsychological study are justified, I consider a similar study performed by Peter Fenwick, Stephen Galliano, Mary Anne Coate, Vicky Rippere and Diane Brown.[5] They examine persons who claim to be, what Fenwick and his colleagues called, psychic sensitive and who have psychic gifts. Psychic gifts include telepathy, clairvoyance, clairsentience, precognition, intuition, psychokinesis, mystical experience, mediumship, déjà-vu, experiences of past lives, poltergeists, etc.[6] The background of their study contained several neuroscientific research reports that associate religious and psychical experiences with non-dominant temporal lobe dysfunction. This is also the case in epilepsy and other temporal lobe disorders.[7]

5 Fenwick et al. (1985).

6 Psychic gifts (or psychic experiences) seem to be a mixture of psychical experiences (chapter two), religious experiences and experiences of poltergeists, UFO, etc. To Persinger, as mentioned above, all these experiences are religious experiences.

7 As we shall see in the next section, Fenwick et al.'s neurological study of mystical experiences did not confirm the following conclusions drawn by previous neuroscientists: that non-dominant

It is interesting how their research is performed, for instance, which kind of questionnaires did Fenwich and his colleagues use? The participants were 17 students who were known to be psychic sensitive as mentioned above and 17 church-going students who served as a control group for religiosity (not psychic gifts) and attended church regularly, i.e., at least once per week throughout the year. The church-going students were selected by the local Anglican and Methodist ministers. Fenwick and his team maintained that the control group was not ideal, because although the group did have fewer psychic gifts, religiosity may also be associated with non-dominant (right) temporal lobe dysfunction. The first questionnaire contained 67 questions regarding the medical history of the participants, their family history, psychic gifts, head injuries and mystical experiences. Furthermore, the following tests were also given: (1) A shortened Wechsler Adult Intelligence Scale (WAIS) which is an intelligence test. The shortened WAIS included tests for comprehension, similarity, vocabulary and performance subscales, block designs and object assembly. (2) The Rey-Osterreith Test tests the non-dominant temporal lobe functions. This test is a visual memory test in which the participants had to copy, and 40 minutes later reproduce, a complex figure. (3) The Wechsler Logical Memory Test is a verbal memory test for dominant lobe function. The participants had to listen to a story, then repeat it from memory immediately, and again one hour later. (4) The Benton Visual Retention Test was given as a general screening test for cerebral pathology. In this test, simple line diagrams have to be drawn immediately after being shown for only 10 seconds. All these tests are official (neuro)psychological tests. My question, then, is why did Persinger not use at least one of these? Since he relates both religious experiences and epileptic seizure experiences to extra-normal activity of the temporal lobe, why did he not include the Rey-Osterreith Test, which tests the non-dominant temporal lobe functions, and/or the Wechsler Logical Memory Test which is a verbal memory test for dominant lobe function?

---

temporal lobe (i.e., mostly right temporal lobe) pathology particularly in relation to epilepsy is linked to mystical experience, Geschwind (1978), that non-dominant temporal lobe pathology is linked to religiosity, Bear/Fedio (1977) – that temporal lobe epilepsy is linked to sudden religious conversions (however there was no hemisphere preference), Dewhurst/Beard (1970) – that temporal lobe pathologic (TLP) interictal psychoses is linked to religious and mystical states (however did not indicate laterality), Slater/Beard (1963) – that temporal lobe epilepsy (TLE) is linked to intense mystical experiences, Cirignotta et al. (1980) (they reported one case of intense mystical experience during a non-dominant TL seizure). Nelson (1970) found that 10 out of 12 mediums had TL EEG abnormalities. Neppe (1980) found significant similarity between TLE symptoms and paranormal experiences. Sensky/Fenwick (1982), Sensky (1983) and Sensky et al. (1984), however, found that TLE patients who had psychic and mystical experiences had a history of psychiatric illness and that epileptic persons were no more likely to have such experiences than the general population. See further, Fenwick et al. (1985) 35–36.

Another problem is that Persinger did not use any neuroimaging technology to study the brain of the meditating persons as did Newberg and d'Aquili and Azari, and he did not refer to any such study in his research report. This implies that he cannot account for which structures and functions of the brain correlate with different experiences during a meditation session as, for example, Newberg and d'Aquili can. So, how could he conclude that religious experiences are caused by above-normal neural activity in the temporal lobe? If he had used some imaging technique for studying the experiences the meditating participants reported that they had, he would have seen which parts of the brain correlated with their experiences and he would have been able to compare this with the parts of the brain that correlate with epileptic seizure experiences. I will return to this later. However, Persinger would still not have been able to say that this altered activity was caused by cognitive kindling, because this cannot be shown by neuroimaging technology alone. One can only see the neural activity itself, not *how* this neuroactivity was caused.

I believe that I have been able to show that Persinger's method is invalid due to the PPI he used. Fraser Watts comments for instance that "the findings of a correlation between the two sets of signs [seizure experiences and religious experiences] is almost worthless scientifically, because the two sets of questions on which it is based [seizure experiences and religious experiences] overlapped so much that it was inevitable that a correlation would be found between the two".[8] Persinger used a wrong (PPI) or an incomplete method. Furthermore, because the PPI does not measure what it purports to measure, i.e., different experiences that meditating persons regularly have during their meditation, but measures epileptic seizure experiences, this study is also irrelevant.

4.1.2 Gender and Religious Experiences

In Persinger (1997) and (2003), he wants to investigate the implications of religious experiences for the behaviour of men and women. His conclusions are that right-handed women (and left-handed men) have more frequent sensed presence experiences than right-handed men (and left-handed women) and that right-handed men (and left-handed women) interpret their experiences of sensed presence as something alien to the self. Furthermore, right-handed men (and left-handed women) who have such an experience may interpret it as coming from an omnipotent, eternal, omnipresent, extracorporeal being (God) and may use this experience as a justification for

8 Watts (2002b) 82.

their aggression. This aggression could include killing other people, because they could have understood the religious experience as God telling them to do so. Persinger concludes that religious persons are potentially dangerous.

One problem we encounter in this study is that he uses a small one-sided sample to represent the population of study, namely males and females who had religious experiences. The sample only consists of 50 right-handed male and 50 right-handed female persons. There is, for instance, no left-handed male or left-handed female control group. In order to restore this incompleteness, Persinger refers to another neuroscientific study by Gordon,[9] who "suggested that left-handed men share greater commonality with normal right-handed women, whereas left-handed women are more similar to higher functioning right-handed men".[10] Persinger seems to agree with Gordon and concludes that left-handed men should more frequently have religious experiences than right-handed men. Hence, on the basis of this, he concludes that he does not need a left-handed control group in his own study: it suffices to refer to Gordon's. This, I argue, was not a preferable conclusion.

Another problem, besides the fact that the sample may be considered too small and that Persinger does not study left-handed men and women, is that all participants of the experiment were between the ages of 18 and 25. Persinger does not consider younger or older persons. Neither does he consider people from different educational backgrounds (all participants were first-year undergraduate psychology students), nor does he consider different social or religious backgrounds (as a matter of fact he does not mention any). Hence, it is uncertain whether the sample meets the requirements of being a representative sample of the population studied, i.e., men and women who had religious experiences.

Yet another problem concerns, once more, the PPI that Persinger uses, which is similar to the one he used in the study mentioned above on the relation between religious experiences and epilepsy. It contains questions related to what epileptic persons experience during their seizures (CEPLS), even if he may have added some specific questions for the purpose of the study. This means that the PPI test could be considered wrong or at least insufficient. Therefore, I question the relevance of having questions based on the descriptions of epileptic seizure experiences for the study of the implication of religious experiences on the behaviour of men and women.

Since Persingers enquiry concerns the implications of religious experiences for the behaviour of men and women, his enquiry concerns the attitudes of persons having religious experiences. Some important remarks can be put forward.

---

9 Gordon (1991).
10 Persinger (2003) 13.

Richard T. LaPiere distinguishes between four levels of human behaviour: "(1) overt-symbolic; (2) overt-nonsymbolic; (3) covert-symbolic and (4) covert-nonsymbolic" behaviour.[11] The overt-symbolic behaviour includes speaking, writing and gesturing. The overt-nonsymbolic behaviour covers acts such as driving a car or closing a door. The covert-symbolic behaviour refers to thought and the covert-nonsymbolic behaviour refers to feelings and emotions. According to LaPiere, there need not be correlations between the four levels of human behaviour in a given situation. Nevertheless, he writes, "such a relationship is assumed in current interpretation of measured attitudes".[12] Furthermore, according to him attitudinal questionnaires present a symbolic situation, more precisely, the measured attitude is "an overt-symbolic response to symbols which represent a situation".[13] An important issue here is that such an "overt-symbolic response to a symbol may be quite different from the overt-symbolic response to the person or object which that symbol represents".[14] Most of the attitudinal questionnaires do not present symbolizations of *actual* situations. Rather they present symbolizations of *abstract* situations. For example, the question begins with *what would you do if..., or* to relate to Persinger's enquiry: *Would you kill if God asked you to do so?* Hence, were the actual situations to occur, it is not obvious that the informants would actually act in accordance to their responses to the questionnaire. Hence, even if a person has answered *yes* to the questions *would you kill if God asked you to do so?* he or she may have a complete different attitude when actually confronted with the person(s) he or she is supposed to kill.

Another interesting remark LaPiere makes is that since a measured attitude refers to an overt-symbolic response to a symbolization of an abstract (and not actual) situation, it is more correct to say that what is measured are ideologies and not attitudes. However, studies of ideologies can be objective if the data on ideologies are correlated with data from other and relevant social facts. Persinger ignores, or does not find it necessary, to correlate the data from his informants with other relevant data. As mentioned above, he does not consider people from different educational backgrounds nor does he consider different social or religious backgrounds. Furthermore, he did not, as is suggested by LaPiere, compare the answers to actual behaviour. Nor did he compare the implications of religious ideologies with implications of other non-religious ideologies (for example political ideologies), on human behaviour. Hence, Persinger's study is not sufficient to be able to conclude that right-handed men (and left-handed women), are poten-

---

11 LaPiere (1938) 176.
12 LaPiere (1938) 176.
13 LaPiere (1938) 176.
14 LaPiere (1938) 176.

tially dangerous when having a religious experience (i.e., would kill when it seemed to them that their God asked them to do so).[15]

Persinger also uses the Transcranial Magnetic Stimulation apparatus (TMS), which as mentioned in the analysis, is a modified motorcycle helmet embedded with four sets of solenoids arranged along the temporal plane. This method is used to experimentally test whether right-handed men experience the sensed presences as something alien to the self whereas the right-handed women do not, and also whether right-handed women report such experiences more frequently than do right-handed men.

As mentioned in the former chapter, Persinger concluded that his hypotheses were confirmed. This experiment seems to be valid in that it measures what it purports to measure. However, it is questionable whether it measures what Persinger apparently thinks it measures, namely religious experiences. According to the table of results, what can be triggered by TMS are experiences such as sensed presence, vibrations, out-of-body experiences, experiences of spinning around, etc. which may in some cases be described by the experiencer as a religious experience, but which are not typically religious experiences, i.e., doctrinally and soteriologically central to the religions (and their continuities). Furthermore, as mentioned in the analysis, when the frequency and intensity is too high, these magnetic fields may induce seizures.[16]

It seems that Persinger's method is, once again, invalid. It is even difficult to understand why he used the PPI in this particular study as it has nothing to do with epilepsy whatsoever and cannot be representative for the population studied, i.e., men and women having religious experiences. This makes this study also irrelevant. Furthermore, it is questionable whether the TMS measured what it purported to measure, namely the implications of religious experiences for the behaviour of men and women. Because of this, I argue, the scientific validity of his study is questionable.

4.1.3 Vectorial Hemisphericity

The last study I shall evaluate on its scientific validity is the one in which Persinger used the vectorial hemisphericity model along with Transcranial Magnetic Stimulation, Persinger (1993c). He also refers to PET-imaging reports of the brains of epileptic and schizophrenic persons. The result of

15 For further reading on the significance of measuring behaviour see LaPiere (1938). Unfortunately I have not been able, whitin the scope of this work, to dig deeper into the problems of psychological investigations by way of questionnaires. Nevertheless I found it meaningful to put forward some, which I believe to be, important remarks.

16 The Royal Australian and New Zealand College of Psychiatrists (2003) point 7.

the experiment is that if the right hemisphere is stimulated the person has an experience of a sensed presence and experiences anxiety. When the left hemisphere is stimulated the person does not experience any fear. When an intercalation occurs in both hemispheres the person will still experience a sensed presence but the fear will have disappeared. This finding led Persinger to conclude that if the neural processes of religious experiences were understood and hence could be generated experimentally, this could be a powerful tool in the treatment of depression, anxiety, and dissociation. Is this study scientifically valid?

We saw that since "[m]ultiple variants of sensed presence often precede mystical and religious experiences that are frequently followed by sudden, permanent changes in the self-concept",[17] Persinger wants to find a way to understand and isolate the neural processes that generate religious experiences. The methods he uses are relevant in the following way. The vectorial cerebral hemisphericity explanatory model maps the neural structures and functions of the human brain. PET-imaging technology shows the neural activity of the brain. By way of Transcranial Magnetic Stimulation (TMS) of certain parts of the brain, certain experiences can be triggered.

However, the question is whether these methods are relevant for what Persinger wants to measure, namely, the neural activity that generates religious experiences. Firstly, the map (vectorial hemisphericity) is a general tool for guiding one in the complex network of neural structures and functions of the human brain similar to a map of Copenhagen, which is a general tool for guiding us in the complex network of railroads, roads, highways and waterways of the capital of Danmark. It is difficult to see where on the vectorial hemisphericity map different religious experiences are located. Secondly, the PET-imaging could have been of importance if Persinger had used this technology (as Newberg and d'Aquili did) to scan neuroactivity during religious experiences, which he did not. Instead he used the clinical reports of the results of such PET-imaging of persons undergoing epileptic seizure experiences and persons experiencing schizophrenic hallucinations.[18] Thirdly, the experiences triggered by way of TMS are, for example, feeling sensed presences, vibrations, out-of-body experiences, etc.[19] These experiences may be, but are not typically religious or mystical. None of the methods separately or together gives us adequate information concerning which neural activity could correlate with religious experiences.

17 Persinger (1993c) 915.

18 Such experiences, as will become clear below, should not be confused with religious experiences (even if epileptic and schizophrenic persons may, of course, have religious experiences as well).

19 Persinger does not list the different possible experiences in this study, however, they were listed in the study concerning the implications of religious experiences on men and women.

Furthermore, it is not clear what Persinger studies, because there are no distinct hypotheses (there are many assertions which are formulated as hypotheses), nothing is known about the participants, there are no tables showing the results of the TMS experiment and there are no distinct results or a clear discussion. Yet Persinger emphasises the importance of finding the neurological processes that generate religious experiences because this knowledge could become a powerful tool in the treatment of psychiatric diseases.

Persinger's study was repeated in Uppsala by Pehr Granqvist, Mats Fredrikson, Patrik Unge, Andrea Hagenfeldt, Sven Valind, Dan Larhammar, and Marcus Larsson.[20] However, the Uppsala team failed to replicate Persinger's results. They did not succeed in provoking any kind of such experiences. One of the reasons was that Persinger did not use a double-blind study. In the double-blind study, which was used by the Uppsala team, neither the participants nor the experimenters were aware of experimental group assignment. This is an important issue because, according to Granqvist, considering people's general suggestibility to induction (i.e., placebo effects), selective experimenter interaction across conditions could easily lead participants in the experimental condition to have some of the vague experiences, whereas those in the control condition (no magnetic field activated) would be less inclined to have them. To clarify, neither the experimenters in the Uppsala team who had to assist the participants, nor the participants themselves, knew when the TMS was on or off. This was not the case with Persinger's team.

On the basis of the above, I argue that the scientific value of the study is questionable. In my view, the vectorial cerebral hemisphericity explanatory model, the PET-imaging as used here, and the TMS experiment do not measure, not even together, what they purport to measure.

4.2 Religious Experiences and Epilepsy

After having evaluated Persinger's neuroscientific studies for their scientific validity, we are now ready to further evaluate Persinger's study of the relation between religious experiences and epilepsy. As mentioned in the analysis earlier, Persinger maintains that if we could find the neural substrates of epileptic seizures, we would find those that generate religious experiences.[21] He is not the only neuroscientist who draws the conclusion that epilepsy and religious experiences are somehow related. On the con-

20 Granqvist et al. (2005).
21 Persinger (1987) 17.

trary, this idea is, as we have mentioned earlier, the most widely canvassed neurological theory of religious experiences. Hence we shall see whether, and if so to what extent, other investigations on the relation between religious experience and epilepsy could support Persinger's studies.

As mentioned in the analysis of Persinger's studies, Fenwick and colleaugues have also done research on the relationship between epilepsy and religious experience. The aim of the study is to find out how head injury and cognitive dysfunction and *psychic* experiences are related. Also, as mentioned earlier, their study concerns a group of people who have *psychic gifts* which are compared with those who do not (or very seldom) have them but who are religious. We learned that Fenwick et al. use different methods and tests to achieve their goal. One of their results is that mystical experiences are presented in both groups of participants. However, "[the occurrence of mystical experiences] was not significantly related to non-dominant [right] temporal lobe dysfunction, as suggested by the literature".[22] This literature, i.e., research reports of different neuroscientists, "suggests that the non-dominant temporal lobe is involved in the genesis of the mystical experience and that head injuries, brain pathology, and epilepsy are related to the occurrence of both mystical experience and psychic gifts".[23] This is so, because TLE seizures are associated with above-normal neural activity in the right (non-dominant) temporal lobe. Hence Fenwick and his team did not confirm these suggestions, but they maintain that the persons whom they called the *sensitives* and who had the psychic gifts had significantly more major head injuries. However, there is no definite temporal relationship between the head injuries and psychic (including religious) experiences.[24] However, Fenwick, having performed a more accurate neuroscientific study than Persinger, questions the literature stating that there is a relationship between (temporal lobe) epilepsy (or schizophrenia) and religious experience, and thus they could not support Persinger's conclusions on the relation between epilepsy and religious experiences.

Jeffrey Saver and John Rabin also consider the relationship between epilepsy and religious experience. They want to draw our attention to the distinction between "epilepsy-related religious experiences of ictal, subacute postictal, and chronic interictal occurrence".[25] What they mean is that we should distinguish between what is experienced during the seizure, directly after the seizure and between seizures. The former experiences are the auras, for example, feeling detached from others and the environment,

---

22 Fenwick et al. (1985) 42.

23 Fenwick et al. (1985) 36.

24 What they do not explain is why if, to give an example, clairvoyance is associated to head injury, it is often the case that what the experiencer sees also happens.

25 Saver/Rabin (1997) 500.

sensation of fear, double consciousness (mental diplopia), which is an experience of one's own consciousness and simultaneously of a consciousness with a different perception of reality, depersonalization, autoscopi (seeing one self), etc. Saver and Rabin maintain that epileptic persons who have a religious background (but not others) will interpret such seizures experiences in religious terms. However, they speculate that even though many epileptic persons do not give their seizure experiences a religious content, it "seems likely that repeated, intense, visceral experiences of the self or the external world as unreal would tend to foster a belief in a supernatural ground of reality and a religious outlook".[26]

Concerning postictal experiences (the experiences succeeding the aura), Saver and Rabin refer to a study performed by Dewhurst and Beard who maintained that six of their epileptic patients experienced dramatic and often lasting religious conversions in the postictal period, i.e., the period following immediately after an exacerbation in epileptic seizure activity, known as the period of *postictal psychosis*. It is maintained that it is more common that postictal psychoses have a religious content.

Concerning interictal experiences (the experiences between seizures), Saver and Rabin refer to TLE persons who show traits of hyper-religiosity and have intense philosophical and cosmological concerns. However, the suggestions that temporal lobe epilepsy and hyper-religiosity are related are not generally confirmed. Rather, Saver and Rabin suggest that several mechanisms may, on their own or combined, lead to the development of religious interests. Saver and Rabin, however, do not deny that religious experience may occur during or after epileptic seizures. In this area they support Persinger.

However, since religious experiences are not and cannot only be related to epileptic seizure experiences, Saver and Rabin suggest a *limbic marker* to be the core element of religious experiences, more precisely numinosa, providing "the sense of having touched the ultimate ground of reality and the sense of the unutterability or incommunicability of the experience".[27] The term *Numinosum* derives from Rudolf Otto and means the sense of awe of an experience of God or Ultimate Reality. According to Otto, the numinosa includes both the element of fear and trembling in the presence of God or Ultimate Reality *mysterium tremendum* and the *fascinosum* of the love coming from God or Ultimate Reality.[28] It is maintained that a numinosum experience occurs spontaneously, i.e., independent of the experiencer's will. However, according to Newberg and d'Aquili, numinosa experiences may

26 Saver/Rabin (1997) 503.
27 Saver/Rabin (1997) 507.
28 Otto (1971) 8–41.

be obtained through meditation.[29] Saver and Rabin argue that similar to strong emotions such as the emotion of being in love, which correlates with increased limbic activity, these religious experiences cannot be communicated entirely. They maintain that their limbic marker theory can be tested by neuroimaging technology. According to Saver and Rabin, "functional neuroimaging during numinous experiences in individuals who have repeated religious transports would reveal alterations in limbic system activity".[30] Furthermore, they maintain that even if the neural substrates of such religious experiences are determined, it does not diminish their spiritual significance. Hence, Saver and Rabin do not support Persinger's view that there is a significant relation between epilepsy and religious experiences. Rather, they associate religious experiences with limbic activity.

Michael Trimble, who also studied the relation between religiosity and epilepsy, refers to the uncertainty of the suggestion that religious experiences are related to epilepsy. He maintains that only six to seven percent of epileptic persons manifest religious experiences. Furthermore, according to him it is not the case that epileptic persons have religious experiences all the time as Persinger maintains.[31]

Hence, it seems that support for Persinger's suggestion that if we were to find the neural substrates of epileptic seizures we would find those that generate religious experiences, is rather weak.

4.3 Comparing Religious and Epileptic Experiences

The question I want to raise in the present evaluation follows naturally from the conclusion drawn above, namely, whether or not religious experiences and epileptic seizure experiences can be compared in this way. Fraser Watts draws our attention to "[a] clear distinction [that] needs to be made between religious experiences *while* people with TLE are undergoing seizures, and experiences and preoccupations *between* seizures".[32] He argues that if an epileptic person has a religious experience under non-seizure-conditions, then if he or she has religious experiences during his or her seizures it is not that sensational. Similarly, if an epileptic person is very depressed during non-seizure-conditions, then if he or she experiences anxiety during his or her seizures it is not sensational either.

---

29 As we shall see when analysing the work performed on religious experiences by Newberg and d'Aquili.

30 Saver/Rabin (1997) 508.

31 My notes from the international symposium on *Epilepsy, Neurobiology and Religious Creativity* in Vadstena, Sweden, 20-22 March (2003).

32 Watts (2002b) 81.

Watts maintains also that seizure experiences may sometimes appear to be like religious experiences but the religious aspects of these seizure experiences are in many ways unlike religious experiences. For instance, he maintains that seizure experiences are often laden with anxiety, while religious experiences in contrast have positive emotional consequences. He writes that "[i]t is a standard feature of powerful religious experiences that they result in a positive emotional tone, even when people have been severely stressed before the experience".[33]

Unfortunately, even if religious experiences are often associated with positive emotions this is not a standard case. As mentioned earlier, according to Carl Gustaf Jung, religious experiences can be so powerful that they destroy the experiencer totally (both mentally and physically) if he or she does not have the adequate religious framework to encapsulate the experience. As an example of such an anxiety laden religious experience, Jung uses the religious vision experienced by a Swiss mystic by the name of Brother Klaus. It is told that he was so terrified by his vision that his face was deformed in such a way that people who met him afterwards were frightened:

> All who came to him were filled with terror at the first glance. As to the cause of this, he himself used to say that he had seen a piercing light resembling a human face. At the sight of it he feared that his heart would burst into little pieces. Overcome with terror, he instantly turned his face away and fell to the ground. And that was the reason why his face was now terrible to others.[34]

Brother Claus recognised his experience as a religious one but was frightened to death when he was confronted with it. The experience literally deformed his face so terrified was he. The case of Brother Klaus is not unique. For example, also Saint Paul's experience on the road to Damascus was a terrifying experience for him.[35] Furthermore, both the Scriptures and other literature describing religious and mystical experiences show that fear, at least in the beginning stage of the experience, is not unusual.

Both Brother Klaus and Saint Paul were religious persons and were acquainted with religious experiences. Concerning Brother Klaus, it was only after many years of contemplation that he concluded that what he had seen was the Holy Trinity and he painted his vision on the walls of his room. Jung writes that "[t]hrough subsequent elaboration this vision then became the so-called Trinity Vision".[36] Hence, anxiety cannot be the locus of difference between religious experiences and epileptic seizure experiences.

---

33 Watts (2002b) 81.

34 Jung (1958) 319, the whole story can be read on 316–323.

35 KJV (1984) Acts, 9:3–18

36 Jung (1958) 319. Brother Klaus (Nicholas of Flüe) was canonized by Pope Pius XII in 1947. The painting can be seen in the church in Sachseln.

In order to clarify what, in my view, could be the defining difference between the two types of experience, I shall present a passage from Fédor Mikhailovich Dostoevskij's novel *The Idiot* as an illustration of an epileptic seizure experience and compare it with a religious experience witnessed by Beardsworth.

A strange light illuminated his heart and mind. All agitation was calmed, all doubt and perplexity resolved into a superior harmony, a serene and tranquil gaiety, which yet was completely rational. But these radiant moments were only a prelude to the last instant – that was immediately succeeded by the attack. That instant was, in truth, ineffable. [...] What matter, after all, though it be a disease – an abnormal tension – if the result, as I with recovered health remember and analyze it, includes the very highest degree of harmony and beauty; if at this moment I have an unspeakable, hitherto unsuspected feeling of harmony, of peace, of my whole nature being fused in the impetus of a prayer, with the highest synthesis of life? [37]

It is known and clinically documented that the famous Russian writer Dostoevskij suffered from epilepsy. He also used his own experiences and transferred them to the characters in his novels. One such character is Prince Mysjkin in *The Idiot*. The prince sees a light and remembers it after his seizure as the highest synthesis of life or source of all life.

Beardsworth writes:

During the night of September 9th 1954, I awoke and looking out of my window saw what I took to be a luminous star which gradually came nearer, and appeared as a soft slightly blurred white light. I was seized with violent trembling, but had no fear. I knew that what I felt was great awe. This was followed by a sense of overwhelming love coming to me, and going out from me, then of great compassion from the Outer Presence. After that I had a sense of overpowering peace, and indescribable happiness.[38]

Both experiences are described as experiences of light, of peace, harmony and happiness. Both are described as experiences of something Ultimate. Hence, from the description of the experiences alone we cannot determine whether we are dealing with a religious experience or an epileptic seizure experience (or both). How then could neuroscientists determine whether both experiences are religious ones or epileptic seizure experiences or whether one is a religious experience and the other an epileptic seizure experience? As mentioned above, Persinger did not perform any neuroimaging on the brains of the participants of his investigation during deep meditation. If he had, he would have paid attention to the neurological differences that underlie religious experiences and epileptic seizure experiences. For instance, Persinger would have taken account of the importance of the role of the prefrontal cortices for both religious experience and epileptic seizure ex-

37 Dostoevskij, cited by Arts (2001) 299.
38 Beardsworth, as quoted by Alston (1991) 18.

periences. According to Scott Atran, brain imaging studies have shown increased neural activity in the right temporal lobe area during epileptic seizures and schizophrenic hallucinations, and religious experiences, such as speaking in tongues, experiences during a trance state, deep meditation and prayer. However, whereas the scan-image showed decreased activity in the frontal cortices during the schizophrenic hallucinations and the epileptic seizures (as a consequence of decreased awareness of reality), it showed increased neural activity in the frontal cortices during meditation and prayer (due to increased awareness of reality).[39]

Hence, it seems that religious experiences and epileptic seizure experiences have different neuroanatomies and that comparing these experiences would be similar to comparing a cold with hay-fever. The symptoms of a cold are very similar to the symptoms of hay-fever, such as, snivelling, but a cold and hay-fever are completely different things. A cold is a virus while hay-fever is an allergy type. Their correlating neural activity is different, which is also the case for religious experiences and epileptic seizure experiences. The symptoms of the religious experience can be very similar to the symptoms of an epileptic seizure experience, such as experiencing a light. However, the correlating neural activity is different.

Another important remark concerning the difference between religious and epileptic seizure experiences is that religious experiences are seen by the experiencers as the results of a communication with God or Ultimate Reality while seizure experiences might be seen as such by the epileptic person but usually are not.[40]

Also the neuroscientist Anne-Marie Landtblom who took part in the discussion concerning epilepsy and religious revelations with a focus on Saint Birgitta of Sweden, puts her finger on some differences between religious experiences and epileptic seizure experiences.[41] Why Saint Birgitta is assumed to have had epilepsy is because Carl Herman Hjortsjö examined her skull (1952) and raised the question whether she may have had epilepsy.[42] However he never wrote down his statement. Birgitta had a tumor at the very top of her forehead, overgrowth of bone at the inner table of the frontal bone (hyperostosis frontalis interna), something that mostly affects older women.[43]

---

39 Atran (2002) 147–166.

40 Schizophrenic hallucinations are the results of the death of certain brain cells in schizophrenic persons. However, epileptic (and schizophrenic) persons can have religious experiences as a result of Divine communication in the same way that a person who suffers from hay-fever simultaneously may have a cold.

41 Anne-Marie Landtblom works clinically with epileptic persons. My notes from the international symposium on *Epilepsy, Neurobiology and Religious Creativity* in Vadstena, Sweden, 20-22 March 2003. See also: Landtblom (2000).

42 Landtblom (2000) 6044–6045.

43 Landtblom (2000) 6045.

Neuroscientists maintain that this disease might have caused epileptic seizures as such tumors can become very big and multiply. If, Hjortsjö and his colleagues are correct concerning the location of the tumor, and if it were the case that Birgitta had epilepsy, then, according to theologian and medical doctor, Åsa Lowén, she should have had *frontal lobe epilepsy* and not *temporal lobe epilepsy*.[44] Epileptic frontal lobe seizures have, as one of their hallmarks, that the seizures are accompanied with cramps or complex automatisms, and rarely with aura. These automatisms are many and often grotesque and hence hard to elude observation.[45] Knowing that only part of her skull has been retrieved, my point is that the statement that Saint Birgitta suffered from epilepsy is founded upon speculation.

According to Landtblom, Saint Birgitta experienced a commonly mentioned epileptic symptom, the *glowing globe* that she described as the Mother of God talking to her from a glowing globe. Landtblom says that this is one of Saint Birgitta's experiences that is sometimes considered as an epileptic seizure experience. However, according to Mogens Dalby, the concept *glowing globe* is not an internationally known denotation for epileptic seizure phenomena. Sometimes, epileptic persons remember their seizure experiences as having seen a light.[46]

Landtblom continues that there are many facts that speak against Saint Birgitta having had epilepsy. For instance, the revelations of Saint Birgitta were too many, too detailed and too varied to belong to an epileptic person. Epileptic persons commonly experience the same experience over and over again. Furthermore, their experiences are stereotyped, which was not the case for those of Saint Birgitta.[47] Furthermore, the revelations of Saint Birgitta included a high amount of information.[48] According to the epileptologist Walter van Emde Boas, what epileptic persons write in their diary is merely what they have eaten and done that day. There are seldom, if ever, spontaneous writings about their inner life. They often use short and simple sentences.[49] Saint Birgitta, on the contrary, wrote down all her experiences carefully. Yet another arguement that renders doubts on the suggestion that Saint Birgitta had epilepsy is that there are no descriptions of her having

---

44 Åsa Lowén was a colleague of mine at the Department of Theology, at the Uppsala University.

45 Mumenthaler/Mattle (2004) 527–529.

46 Dalby suggests that the term glowing globe may have come into the medical language as follows: A person tells a doctor about his or her seizure, that he or she saw a light. The doctor asks: "like glowing?" The patient would answer: "yes, glowing and turning." The doctor: "Like a glowing globe?" Yes, the patient says just for the sake of the recognition of his or her vision. The doctor will ask the next patient if he has experienced some kind of glowing globe? Consequently, the glowing globe becomes a term used in medical contexts.

47 Landtblom (2000) 6046.

48 Bergquist et al. (2003).

49 My notes from *Epilepsy, Neurobiology and Religious Creativity*: Internationellt symposium i Vadstena, Sweden, 20-22 March (2003).

seizures. This is strange, because there is so much written about her and epilepsy was not considered to be a mental disease as it is today; rather it was often considered a gift from God. However, I believe that one cannot completely exclude the idea that she might have had epilepsy. Dostoevskij suffered from it as well and yet his literary work belongs to the best in the world. I believe that there are not only different types of epilepsy but also different degrees within each type. Furthermore, most epileptic persons live a completely normal life just as you and me. Only few are hospitalized or institutionalized. Generally, these persons' lives are very dramatic and not many survive a long life of activity in contrast to Saint Birgitta who lived in the beginning of the 14th century and lived to be 80 years old.

Persinger also concludes that religious experiences and epileptic seizure experiences are related to each other in the sense that both are caused by kindling the temporal lobe. However, we saw that cognitive kindling refers to a neuroscientific method (used in the research on epilepsy) that provokes epileptic seizures by way of stimulating specific parts of the brain, such as the temporal lobes, with weak electrical or chemical stimuli.[50] For instance, Leslie Brothers performed such an experiment on epileptic persons. According to her, stimulating the temporal lobes of epileptic persons caused perceptual experiences, which can be visual hallucinations (seeing a face), auditory hallucinations (hearing a voice or music), or hallucinations concerned with other sensations as, for example, taste or smell. The subjects experienced fear when their amygdala was stimulated. The person might turn pale, look frightened or even scream. Some may also experience reliving something familiar – as if they have been there before, déjà-vu.[51]

Even if religious experiences may sometimes provoke anxiety, it is questionable whether they can be caused by cognitive kindling. Persinger perhaps associates religious experiences with cognitive stimulation because one characteristic of meditation is to stimulate the brain in a specific sense such as through focusing on an image or mantra. But that is not kindling, nor is this sensational. There are many kinds of brain stimulation, writing this book stimulates me, as does being with the persons I love or listening to a piano concert. Persinger blends the concepts of kindling and stimulation and it is difficult to understand what, exactly, he means.

Religious experiences and epileptic seizure experiences have a different neurological basis and even if some epileptic persons may have religious experiences, Persinger's statement that if we find what neural activity underlies epileptic seizures, we have found what neural activity underlies religious experiences, is rather doubtful.

---

50 See the analysis above.
51 Wildman and Brothers (2002).

4.4 Right-Handedness of Men and Women

According to Persinger, right-handed men remember religious experiences as an *alien presence* (i.e., an outside body experience). Right-handed women remember their religious experiences as a *union* with, among other possibilities, Ultimate Reality (i.e., an inside body experience). We also saw that, neurologically, left-handed women are similar to right-handed men and should therefore experience alien presences, and right-handed men are similar to left-handed women and hence should experience union with, for example, Ultimate Reality. His conclusion is that right-handed men and left-handed women are potentially dangerous when having a religious experience because they may use these religious experiences which they may interpret as hearing the voice of God, as justification for their aggression, and hence kill in the name of God.

Extreme consequences of Persinger's conclusion would be that (1) all religions of which one of the characteristics is the search for union with Ultimate Reality should have, as their originators, right-handed women and left-handed men and that the religious leaders and religious people belonging to these religions who are able to obtain such a union should be right-handed women and left-handed men and furthermore, that these religious persons should not be considered to be potentially dangerous; (2) all religions that have as one of their characteristics the search for a Divinity should have as their originators, right-handed men and left-handed women and that the religious leaders and religious people belonging to these religions who are able to *experience* that Divinity should be left-handed women and right-handed men and furthermore, that these religious persons should be considered to be potentially dangerous for the reasons given above.

From the point of view of religious studies, experiencing a union with undefined Ultimate Reality is typically associated with Southeast Asian religions (Buddhism, Taoism) while an experience of *a* Divinity (Jesus, YAWH, Allah) is typically associated with the Abrahamic religions (Christianity, Judaism, Islam).[52] If Persinger is right, religious leaders and religious people belonging to the Southeast Asian religions and who have religious experiences should be right-handed women and left-handed men and religious leaders and religious people belonging to the Abrahamic religions and who have religious experiences should be left-handed women and right-handed

52 Of course, this does not mean that believers belonging to an Abrahamic religion cannot describe their experience as an experience of a union with an undefined Ultimate Reality. Furthermore, I suggested that we should understand God or Ultimate Reality as the *absolute ground* of everything that is (and ever will be). Unio Mystica is described by some Christian mystics as an experience of a union with God. Similar believers belonging to Southeast Asian religions may also describe their experience as experiencing the presence of a spiritual being.

men. If he is right, then especially the latter religious group should be potentially dangerous when having religious experiences. Hence, Persinger's conclusions, if correct, could have consequences for religious studies and how the differences between Eastern and Abrahamic religions could be understood.

However, according to Laughlin, what determines where the neural activity will take place (in the right or left hemisphere or in both) is the type of meditation and the accompanying beliefs. Contrary to Persinger, they maintain, for instance, that meditation will generally lead to a more balanced brain processing, i.e., there will be symmetry rather than asymmetry of neural activity in both the hemispheres. Furthermore, they maintain that the different meditation techniques,

> accompanied, as they always are in traditional settings, by cosmological beliefs and mythopoetic symbolism, may result in different neurocognitive entrainments [...] This combining of cultural elements will involve variant entrainment of left and right-lobe [hemispheric] functions.[53]

Hence, Laughlin et al. could not support Persinger's conclusions of religious experiences on the behaviour of men and women. They refer to the importance of the cultural, religious and personal background of the believers for the neural activity that correlates with the religious experience.[54] Moreover, Laughlin et al. emphasise symmetry instead of asymmetry of hemispheric activity instead of dominance of one of them as Persinger does.

Nevertheless, some studies might, at least to some extent, support Persinger's. Consider the following conclusions that Elkhonon Goldberg draws from his research on brain damage.[55] In a neuroscientific study, both healthy male persons and male persons with frontal lobe damage where asked to perform the cognitive task of making decisions. Goldberg discovered that during such decision-making, one of the hemispheres (frontal lobes) of healthy men would become dominant depending on the kind of the decision, i.e., depending on the situation. However, the men did not show any extreme context-independent (right-hemispheric dominant) nor extreme context-dependent (left-hemispheric dominant) behaviour. There was a dynamic balance. This balance, he discovered is lost in male persons with frontal brain damage, i.e., there is extreme context-dependent (left-hemispheric dominant) behaviour. He also compared healthy female persons with frontal lobe damage with women who had none. Female persons with frontal lobe damage seemed to produce extremely context-dependent

53 Laughlin et al. (1993) 149.

54 I shall regard the importance of the cultural-religious and personal background of the believer more carefully when I evaluate the research on religious experience by Newberg and d'Aquili below.

55 Goldberg (2001).

behaviour, whereas healthy women exhibit context-independent behaviour. He concluded that the female frontal lobes function more alike, damaged or not.[56] According to Goldberg this means that

> [t]he expectation is that healthy right-handed males with the preference for context-dependent decision-making will show a particular activation of the left prefrontal cortex while doing the task. By contrast, healthy right-handed males with the preference for context-independent decision-making will show a particular activation of the right prefrontal cortex. [...] Healthy right-handed females with the preference towards context-independent decision-making will show a particular activation of the prefrontal cortex bilaterally, and the females with the preference for context-dependent decision-making will show a particular activation of the posterior cortex bilaterally. The male and female cognitive decision-making strategies are different, and so are the patterns of lateralization of their frontal lobe functions.[57]

Hence, what Goldberg says is that there are biological differences in the male brain (the left-right differences being better articulated) and female brain (the front-back differences being more articulated), both having to do with the cognitive brain structures and functions. This could be understood as a weak support for Persinger to the extent that both Goldberg and Persinger emphasise a significant difference of correlations between cognitive tasks and the hemispheres of men and women. (Religious experience being, at least to an extent, a cognitive issue.)

However, there is a difference in the context of the conclusions drawn by Goldberg and Persinger. Goldberg would not maintain that *all* right-handed men experience religious experiences in the same specific way and similarly, *all* left-handed men, *all* right-handed women and *all* left-handed women, since he emphasises the complex relationship between left-or right-handedness and hemispheric specialization. Persinger does not consider this at all. Hence, the support Persinger could find in Goldberg's research is rather general.

Another study, performed by Ron Philipchalk and Dieter Mueller, might also give Persinger's conclusions some support (even if the support is weak). They performed a neuroscientific experiment with a result that indicated increasing activation of the right hemisphere relative to the left after speaking in tongues, and increasing activation of the left hemisphere after reading aloud. They measured the temperature of the participants before and after the tasks by using sophisticated instruments and found that speaking in tongues may be associated with increased activity in the right hemisphere, which would support Persinger.[58]

---

56 Goldberg (2001) 93–99.

57 Goldberg (2001) 95–96.

58 Specifically, they employed infrared radiation. A sophisticated infrared camera that measures temperature changes. Philipchalk/Mueller (2000) 181–185.

This support is weak for the following reasons. Firstly, Philipchalk and Mueller write that "this [...] in no way suggests that glossolalia [speaking in tongues] is nothing but right-brain activation".[59] This means that they explain religious experiences in terms of *correlations* with the brain rather than in terms of religious experiences being generated by the brain as Persinger does. Persinger, for instance, maintains that speaking in tongues is a consequence of group manipulation by way of, illustratively, the right sort of music. He says that

> [w]hen the entire group is activated, a mob-like psychology appears. At the peak of the group arousal, the members are convinced they are right. [T]here is no rational counterargument. [...] Since everyone's brain is stimulated, even the most nonsensical events become meaningful. [...] Some episodes of "speaking in tongues," [...] show this effect. As the number of brain stimulation increases and the dissociation begins, basic sounds are emitted. Most of them are elementary composites of what the person has heard (or thinks are the sounds of Hebrew, for example). Since the lower parts of the brain are still functioning, the sounds are still integrated phonemes, although the sequences are meaningless.[60]

Persinger does not refer to any research to support his assumptions but we have seen earlier that he maintains that a group of people can be manipulated by way of music, because music stimulates (kindles) the temporal lobes, which in turn trigger religious experiences, which in turn could be used for political purposes.[61]

Secondly, Philipchalk and Mueller recognise that there were many shortcomings in their study, namely, there were only five participants, there was no control group of people who do not speak in tongues, there was no control task in which people performed a more traditional right-hemispheric task, there was no task in which people mimic glossolalia by babbling etc. Hence, the scientific validity of this experiment is, without doubt, questionable and it is clear that more research is needed to be able to draw reliable conclusions. Finally, they suspect that the enquiry form in which the participants had to answer questions such as "When did you first speak in tongues? What was it like? How often did you speak in tongues?" may have supported tongues-related processes, including visualization of past experiences which influences the outcome. Contrary to Persinger, then, they question the questionnaire that was used and the validity of their experiment.

According to Mogens Dalby, the left-right-handedness theory was a popular psychological test during the Seventies and Eighties. The tests were used to help people making choices concerning employment, study and

---

59 Philipchalk/Mueller (2000) 185.
60 Persinger (1987) 105.
61 See the analysis and evaluation of Persinger (1997/2003).

sports. The theory was later found to be misguided, because it did not consider the complexity and plasticity of the brain. It was found to be too much of a black-and-white theory.[62] To give an example of how complex the human brain is let me quote Ramachandran:

> The brain is made up of one hundred billion nerve cells or 'neurons', which form the basic structural and functional units of the nervous system. Each neuron makes something like one thousand to ten thousand contacts with other neurons [called synapses]. [...] [I]t has been calculated that the number of possible permutations and combinations of brain activity, in other words the numbers of brain states, exceeds the number of elementary particles in the known universe.[63]

It seems that there is no significant support for Persinger's thesis that there is a specific correlation between hemisphere dominance and gender, and hence a specific correlation between hemisphere dominance and types of religious experience. I believe that his conclusion that right-handed men and left-handed women are potentially dangerous when having a religious experience is rather doubtful.

4.5 Hemisphericity and Religious Experiences

It is now time to evaluate Persinger's study in which he searches for a way to isolate the neural processes that generate religious experiences, because he believes that the results of such isolation may be important for clinical use. His idea is that God Experiences are often followed by sudden, permanent changes in the self-concept, which as seen in the analysis, may be the reason behind some religious person's sudden recovery. Persinger argues that if we could understand and isolate the neuroelectrical basis of mystical experiences and religious conversions, these bases could be generated experimentally and applied clinically and become a powerful tool for the treatment of depression and anxiety.[64]

We have seen that he uses a theoretical tool called vectorial hemisphericity, which is a kind of map of the brain and its functions, and illustrates three metabolic states of activity of the left and right hemispheres: below-normal, normal and above-normal. When the right hemisphere was stimulated by way of Transcranial Magnetic Stimulation – TMS (until above-

62 The individuals (usually persons who aimed at becoming political leaders, managers, or coaches) had to fill in a questionnaire which would indicate if they were left-handed or right-handed. The answers were ordered into a two-dimensional diagram. The result was brought about by way of a vector. From there stems the name *vectorial hemisphericity* model, which will be evaluated below.

63 Ramachandran (2003) 4.

64 Persinger (1993c) 926–927.

normal activity was reached), some participants reported having had a sensed-presence experience and some reported also having experienced anxiety. I named these experiences, Religious Erlebnis (RErl). This was not the case when the left hemisphere was stimulated by TMS. I called left-hemispheric experiences, Religious Ideological Experiences (RIE). Anxiety was also not involved when both hemispheres were stimulated simultaneously, (and thus an intercalation occurred). I have named such experiences Religious Erfahrung (RErf), which as it were, combines RErl and RIE, i.e., the RErl has been put in a frame of reference. An example would be a sensed presence experience that is interpreted by a Christian believer as the hand of God. That there was no fear involved was why Persinger concluded that the isolation and understanding of the underlying neural structures of religious experiences can be important in the psychiatric care of clinically depressed persons.

Since pleasant mood is associated with a normally activated left hemisphere and anxiety with an above-normal activated right hemisphere, and since above-normal neural activity in the left hemisphere is associated with high self-esteem, and provided that Persinger's study holds, clinical manipulation of the left temporal lobe might reduce the anxiety, which clinically depressed persons suffer from. Nevertheless, it is hard to believe that the experience that the patients would have during such magnetic wave therapies would turn them into religious persons, and that TMS would become a shortcut to religious belief.[65] For instance, Gregory Peterson asks, "in what sense can we be sure that such experiences are truly of God"?[66] He argues, why should we accept Persinger's claim that the experiences that the subjects of the research had while they were tested by his Transcranial Magnetic Stimulation (TMS) apparatus really were religious experiences?

From chapter two we know that religious experiences are those experiences in which it seems to the experiencer that he or she is confronted with God or Ultimate Reality in such a way that it is in accordance with the doctrines and soteriology of his or her religious tradition. It is doubtful whether the descriptions given by Persinger in his research reports of what the persons experience during the TMS experiment (i.e., vibrations, out-of-body experiences, experiences of spinning around, etc.) would pass the test of being central to the doctrines and soteriology of the religious traditions (and continuities), even if one or two maybe would. Clearly, a person who maintains that she has experienced Jesus does not mean that she or he experienced her temporal lobe receiving secondary visual and auditory trans-

---

65 Which of course does not mean that this could never be the case.

66 Peterson (2003) 116.

missions resulting in a sensed presence, something she or he later described as Jesus. What the person means, in fact, is that she or he experienced more than all that.

Furthermore, assume that it is true that religious people cope with their illnesses and crises in a better way than non-religious people. The reason for this is probably not (for most of them) that they had a God Experience but that they have a strong religious faith which gives their lives a certain purpose. For instance, from the point of view of the Christian religion, it is not (always) meaningless to be ill or to undergo a personal crisis. Sickness and depression may have a specific meaning for the life of the suffering person(s). Such a religious belief cannot, I maintain, merely be obtained through Transcranial Magnetic Stimulation of the brains of clinically depressed persons or other persons with temporal lobe dysfunctions.

There is no clear distinction between the neural activity in the left and right hemisphere. Everything we do or experience is associated with complex neural activity in the entire brain area, even though neuroscientists agree that the right hemisphere is more associated with non-verbal thought, visual spatial perceptions, and the perception, modulation and expression of emotions than the left hemisphere. They agree also that the left cerebral hemisphere is more analytically inclined, and is more associated with language and mathematical processes than is the right hemisphere. Therefore, I suggest that we could accept that religious experiences are associated with the right hemisphere for the part of the experience that concerns, for example, visions, hearing a voice or feelings, and with the left hemisphere for the part of the experience that concerns the understanding and interpretation. However, this does not mean that religious experiences can by default be provoked by TMS or that the neurological substrates that generate them can be isolated, copied and used for medical purpose in the sense that Persinger intends. This is because there are certain norms to which a religious experience has to conform. For instance, it has to be both recognised as such by the experiencer and by the religions through the doctrines and soteriology that are central to them.

4.6 God Experiences and Personal Crises

We shall proceed by evaluating Persinger's idea that experiences of Ultimate Reality are generated by the biology and/or psychology of personal crises. Persinger explains Ultimate Experiences, or God Experiences as he prefers to call them, in terms of a consequence of existential crises, the most severe being the anticipation of one's own death and the death of a loved-one. As seen in the analysis, this is so, because according to Persinger

> [a] biological capacity for the God Experience was critical for the survival of the [human] species. Without some experiences that could balance the terror of personal extinction, existence of the human phenomenon called the 'self' could not be maintained.[67]

This biological capacity evolved along with the evolution of the neocortex. Thus, this capacity is seen as a consequence of the human brain's construction, i.e., should the temporal lobe have developed differently, the God Experience would not have occurred.

At the time of the creation or evolution of the human species, the species was provided with a neocortex. As mentioned earlier, this is what makes the human species human. If the human brain were different and did not have such a developed neocortex, humans would not be able to create language, have a culture, make plans or perform other sophisticated tasks. Fish do not have a neocortex, reptiles and birds have but a poor beginning of it. However, from the premise, if the human brain did not consist of such a developed neocortex, it does not follow that there would not be any God Experiences. A conclusion that could follow from this premise is that the persons having God Experiences would not put them within the framework of their cultural-religious-personal reference, because of the inability to create language, culture, religion, art, etc.[68] This conclusion, I argue, does not imply that there cannot be any God Experiences only that such experiences would be described, as it were, differently by the experiencers.

Persinger maintains that when we undergo deep existential crises and due to biological or psychological factors, i.e., we may be exhausted and/or anxious, we are more prone to have a God Experience. Furthermore, since, due to this God Experience the person may feel reborn, and since to be born again is a soteriological issue in most religions, persons who are not religious from the start may join a church or another religious community. However, according to Persinger, these experiences have nothing to do with God or Ultimate Reality. What happens is rather that the person re-experiences her or his relationship during infancy with her or his parents or parent-substitutes. Persinger concludes that the person erroneously mistakes this experience for a religious one and asserts that religious experiences are illusory.

Remember the story of Veronica who was about to drown but was rescued? According to her, she was saved by God who answered her scream for help. To Persinger, what saved her was the floating log in the water and nothing else. The reason that Veronica believed that God saved her was due

---

67 Persinger (1987) 12.

68 As mentioned earlier, the temporal lobe is located along the sides of the head and is associated with language and conceptual thinking.

to her mixing up her situation with a particular situation in her infancy, where a parent came to her aid after she called for help. Persinger maintains that Veronica confused her parent with God, that she confuses the God concept (saviour, carer) with the parent concept (saviour, carer).[69]

It is, of course, possible that some people wrongly mistake an experience for a religious experience. Persinger is right in this. However, he is mistaken when he maintains that all experiences that people believe to be religious are illusory. Persinger cannot, from the premise that (1) some *x* believes that *y* and (2) it seems that *y* is not *y* but *y*', conclude that (3) all *y* are *y*'.

Persinger is not the only (neuro)psychologist who maintains that there is a relation between personal crisis and religious experiences. Antoon Geels has for many years studied people who have undergone an existential crisis which ended with a religious experience. However, it is doubtful whether Geels' explanations support Persinger's standpoint.[70] For instance, an important difference between the two explanations is that Geels does not (as Persinger does) reduce religious experiences to some psychological or neurological process. On the contrary Geels maintains that (neuro)psychology is only one "way to look at" reality.[71] It does not exclude other ways of looking at reality. Thus, contrary to Persinger who maintains that science can explain religious experiences, Geels holds that scientific explanations are only some of many possible explanations. Another important difference between Persinger and Geels concerns their neuropsychological research. Geels uses open questions in his questionnaires in order to let the persons describe their experiences the way they remembered them. Geels calls this part of his method "Critical Incident Biography" and is meant to be a kind of curriculum vitae focusing on the person's own picture of the religious experience that influenced his or her life in a drastic manner.[72] According to Persinger, religious experiences during severe existential crises are due to "spontaneous intermittent above-normal right hemispheric metabolism [that] could encourage intense brief intrusions that would ultimately kindle or elevate [...] the left hemispheric function",[73] and argues that what the experiencer has to say about his or her experience is not essential. Since left hemispheric functions are associated with language, logic, rationality, etc., "[t]he phenomenological correlates would include enhanced self-esteem,

69 Whether this example concerns a matter of miracle or not is not essential. What is essential is how she explains her experience, she does not say that it is a miracle but that God answered her prayer and how the experience is explained by Persinger.

70 Geels is a psychologist of religion using research methods related to his discipline which are different from Persinger's.

71 Geels (2001) 19, my translation. He calls these ways of looking at reality perspectivism. Geels pleads for a nuanced explanatory model.

72 Geels (2001) 21.

73 Persinger (1993c) 924.

'restructuring of the self-concept'",[74] and sudden understanding. Such an emerging of "the new self" is not unusual and is often accompanied by a most intense religious experience.[75]

A contrasting explanation of the parent-God relationship comes from Ashbrook and Albright. They write that "the warmth of religious devotion, the strength derived from religious trust, confidence in what is to come – all are ways of establishing a 'safe and secure' relationship, not simply with a parent, but with a world, a cosmos".[76] Ashbrook and Albright turn the argumentation the other way round, i.e., instead of projecting, for example, the love and care of the parents to God, the love and care of God is projected to the parents. Understood as such, the safe and secure relationships with our parents or parent-substitutes (and whole environment) are re-experiences of our religious experiences (devotion, trust, love, union, etc.) Which explanation of the God-parent-God relation should be accepted? Persinger's or Ashbrook and Albright's? I suggest that both could be plausible explanations of the parent-God relation. It is not impossible that some project the parental care of their childhood onto the God of their belief and it is certainly not impossible (on the contrary I would say) that some, having experienced Ultimate Reality, become much more careful, peaceful and affectionate towards their environment and thus project the God care and love onto the parents. However, what is not plausible is Persinger's conclusions that all religious experiences are illusory, that any Ultimate Reality or God cannot exist, because religious experiences can be experienced during personal crises.

Are religious experiences illusory and is it the case that no Ultimate Reality or God can exist? Consider the following two examples of a religious experience as they were told to me by one of my informants whom I called Grace:

> One Sunday, when I was about eight years old, I was sitting in the living room on my own, reading the children's Bible. It was natural for me to do so, because I am the daughter of a minister, and I liked the Bible stories very much. Suddenly, a kind of white smoke came out of the page I was reading. Being the daughter of a minister and having heard many Biblical stories, I was not afraid at all but curiously watched what happened. The white smoke turned into figures who smiled at me and left. I remember this experience as something that really happened, and have no doubt about the reality of the spiritual figures.[77]

She continued:

74 Persinger (1993c) 924.
75 Persinger (1993c) 924.
76 Ashbrook/Albright (1997) 101.
77 From one of my informants, my translation from Swedish, cited with permission.

About a year later, on an occasion when I was lying in bed making shadows on the wall with my hands (forming a duck etc.), I suddenly saw that one of my hands burned. There were flames coming from the palm of my hand and my fingers. Again, being the daughter of a minister and having heard so many Biblical stories about fire and since my hand did not hurt at all, I was not really afraid and watched what happened. The fire disappeared as quickly as it came, leaving no marks of course. Later in life, when I became an icon-painter, I remembered my burning hand and believe that this was a Divine sign that my hands would become important instruments for the religious task I would be given.[78]

Are these experiences illusory?

Persinger would say that they are. I suggest that the experience would, according to him, be due to Grace's intensive reading of the Bible (she likes the stories very much hence she read the Bible often) which stimulated her temporal lobes in such a way that an above-normal neural activity occurred which triggered the experience. That she was not afraid would be due to the intercalation of her hemispheres: her right hemisphere being responsible for the experience and her left hemisphere for the rational thinking (her father being a minister).

Persinger would probably explain the second experience in terms of epileptic-like signs, because flames and fire are terms associated with epileptic seizure experiences. He would probably not take into account Grace's description of what she experienced.

Persinger would not take into account that Grace's experiences fulfil all the criteria for an experience to be a religious one, because, as mentioned in the analysis, Persinger maintains that religious concepts are nothing but words, and since we are governed by language in every aspect, we should realise the limitation of language. We should realise that religious phenomena are nothing more than creations of human language. He writes that "God is simply "dog" spelled backwards.[79]

Well, he is correct about that of course, but I wonder if he has considered the opposite, namely that "dog is simply God" spelled backwards? Following his way of reasoning, does this mean that neither gods nor dogs exist? Words, Persinger says, should refer to things that are scientifically measurable. According to him, words which aim at non-empirical referents such as God, eternity, infinity and love, are problematic, because these words do not refer to anything that is measurable by scientific means. Such non-empirical referents then do not exist, except as words, and they can never exist as more than words.[80]

---

78 From the same informant, my translation from Swedish, cited with permission.
79 Persinger (1987) 84.
80 Persinger (1887) chap. 7.

Persinger is not alone in his suggestion that religious experiences are illusory. The first support for his opinion that religious experiences are illusory comes from some neuroscientists who take the medical diagnosis that when the corpus collosum (which connects the two hemispheres) of epileptic persons is cut to prevent them from having epileptic seizures, the religious experiences (epileptic seizure experiences actually) also disappear, as evidence for this assumption.[81]

Since these neuroscientists also consider religious experiences and epileptic seizure experiences to have the same neurological basis, religious experiences have to be a consequence of a malfunctioning of the brain in the same sense that epileptic seizure experiences are.

This support is insufficient, because firstly, the experiences concerned are not religious experiences but epileptic seizure experiences which do not have, as seen earlier, the same neuroanatomy. Hence, from the fact that epileptic persons were released from their seizures by way of brain surgery it does not follow that if the surgery were performed on religious persons, they would not have any more religious experiences. This may be the case but it may also be the case for other human experiences, such as, being in love, feeling hungry and all other experiences that involve the neural activity of both hemispheres.

Secondly, one cannot draw a general conclusion from particular situations. In my view, the fact that some epileptic persons may have a religious experience, does not imply that all epileptic persons have religious experiences. One cannot conclude that all roses are white due to the fact that some roses are. Thirdly, when a person undergoes specific surgery in order to prevent some complaint from occurring, it is obvious and desirable that the complaint disappears. A fourth objection would be that even if our religious experiences were illusory, God could still exist, but does not want to have anything to do with us.

A fifth objection should become clear from the following thought experiment. Assume that a person, for some medical reason, has to have an operation on the auditory structures of the brain in such a way that it takes away the person's hearing capacity completely. Does this mean that the sounds he or she heard before were illusory or that the chirping birds do not exist? Similarly, assume that an epileptic person had a religious experience which disappeared after the operation of the corpus collosum. This does not mean that the religious experiences the person had before the operation were illusory or that God could not exist.

---

81 The first such surgery was performed by the psychologist Roger Wolcott Sperry in 1968 on epileptic patients who had suffered uncontrollable epileptic seizures for a long time. The purpose of the surgery was to reduce the spread of electrical activity and therefore reduce the severity of epileptic seizures or to end them completely. The operation was seriously debated and criticised.

The philosopher Patricia Churchland maintains that Persinger's results do not prove the non-existence of Ultimate Reality, or the illusory status of religious experiences at issue, but they are nevertheless important, because "they drain *probability* from the hypothesis that the experiences provoking God-reports are truly experiences *of God*".[82] Churchland also finds the hypothesis that God exists "improbable at this time. I do believe this not on faith, but on the basis of evidence and argumentation".[83] What she counts as evidence is, for instance, Persinger's studies of the relation between epilepsy and religious experiences and the artificial stimulation of the temporal lobes by way of the TMS (helmet) that *provoke religious experiences*.

However, that they *drain* probability from the hypothesis that such experiences are experiences of Ultimate Reality or God is doubtful if I am correct in my conclusions that we have good reason to question the scientific value of Persinger's research on religious experience. As the results of my evaluation have shown above, his research does not seem to measure what it purports to measure, and is in some cases even biased. For instance, the studies Churchland refers to, Persinger (1993b) and Persinger (1993c) were considered to be irrelevant, because Persinger does not investigate religious experiences but epileptic seizure experiences. It was shown that religious experiences and epileptic seizure experiences have a different neural basis. Hence, these experiences are not related in the way Persinger wants us to believe they are. Moreover, following the table of the results of the experiment with the TMS and what is said above, it is questionable whether the vibrations, out-of-body experiences, experiences of spinning around, etc., the subjects of the experiment underwent can be classified as religious experiences.[84] Churchland is too optimistic about the capacity of science to explain religious experience. She believes that all scientific hypotheses rest on evidence and rational argumentation and that all hypotheses are open to criticism, investigation and to eventually being refuted. She maintains that matters of faith (theology and religious studies) appear to stand above criticism or tests. Science, on the contrary, includes all that she says.[85] Churchland overestimates science and underestimates theology and religious studies. Firstly, because theology and religious studies discuss matters of faith but are not themselves matters of faith. Secondly, matters of faith are not considered to stand above criticism and tests. As a matter of fact, analysing, evaluating, criticizing and testing matters of faith is what

82 Churchland (2002) 387.

83 Churchland (2002) 392.

84 Once more, this does, of course, not exclude the possibility that a person could have a religious experience during such experiments.

85 Churchland (2002) 373–400.

philosophy of religion, a discipline within religious studies and theology, actually does. Churchland's trust in science could be challenged.

However, Churchland could argue that she, in her criticism, does not refer to theology or religious studies but to the religious believers. She is correct in that there are religious believers who do not question their faith but there are also those who do. Hence, her criticism, as a general statement, is unjustified.

Another example of Churchland's overestimation of science and why she thinks that it is improbable that God exists is that she believes that one day science will be able to explain religious experiences. She writes

> When the human scientific community was comparably ignorant of such matters as valence, electron shells, and so forth, [we] could not imagine how you could explain the malleability of metals, the magnetizability of iron, and the rust resistance of gold, in terms of underlying components and their organization.[86]

Thus, in all times, people have not been able to explain certain things, because science does not yet have the capacity to explain them. Surely she is right in this. However, I think it is quite risky to project the past onto the future in such a way. In my opinion, from the historical fact that science, on a certain occasion, could explain why iron swells when heated, something that our ancestors had no possibility of explaining, it does not follow that scientists in the future will be able to explain the origin of religious experiences.[87] They may or may not be able to do so, but there is no logical relation between what science *could* do and what science *will* do. Moreover, even if it should be that scientists one day can explain religious experiences, this does not imply that all religious persons would stop being religious. Furthermore, since we do not know in which manner scientific methods will develop, we cannot know how science possibly would explain religious experiences. Nevertheless, Churchland is wrong if she maintains that scientific explanations are superior to religious explanations.

It is unfortunate that Churchland does not consider other neuroscientific explanations of religious experiences. Neither does she take into account explanations from other disciplines such as theology, religious studies, social and cultural studies, and/or psychology. Furthermore, she does not regard the explanations of the experiencers themselves concerning what they experience. Neither does she consider the different ways in which religious experiences may occur, for example, by way of different meditation techniques or spontaneously, and nor does she consider the wide variety of religious experiences. Churchland's opinion that Persinger's explana-

86 Churchland (2002) 395.

87 Churchland applies eliminating reductionism, a type of reductionism which will be discussed later.

tion of religious experiences drain probability from the hypothesis that experiences of Ultimate Reality really are what they are said to be, is rather problematic.

4.7 Experiencing and Types of Experience

The question is now which ways of experiencing and which types of religious experience Persinger's research may have touched on.

For the sake of clarity, let me briefly repeat what I mentioned in chapter two. An experience (including a religious experience) *x* is an aspect of the subject of experience *y*. Furthermore, an experience consists of *y* and sometimes even of an object *y*', but not necessarily. Being thirsty would be an experience of an aspect of the subject, namely, neural signals from the central nervous system, and the experience of drinking water would even have an object, namely water. We have also seen that there are several ways of (religious) experiencing. These ways are, (1) the *participating perceptive sensory way* of experience, (2) the *participating introspective sensory way* of experience, (3) the *participating introspective non-sensory way* of experience, (4) the *witnessing perceptive sensory way* of experience, (5) the *witnessing introspective sensory way* of experience and finally, (6) the *witnessing introspective non-sensory way* of experience.

It is important to keep in mind the following: (a) *Participating* here means to participate in one's (religious) experience in contrast to *witnessing* one's (religious) experience. It is the difference between participating in a mass as opposed to experiencing witnessing the mass on television. (b) *Introspective* refers to the (religious) experiencing of an aspect of a subject *y* in contrast to *perceptive* which refers to the (religious) experiencing of *y* and an object *y*'. For instance, a person may be experiencing reading the Bible rather than experiencing God's beauty in nature. (c) Finally, *sensory* means that something is experienced using the five senses as against *non-sensory* which concerns thoughts, intuition and feelings. For example, one may experience a religious ceremony in contrast to experiencing Absolute Unitary Being.[88]

What are the religious experiences considered in Persinger's studies? What we know with certainty is that Persinger studies persons who practise meditation, mostly but not only, transcendental meditation. As Persinger

88 Absolute Unitary Being is a term used by Andrew Newberg and Eugene d'Aquili and is considered to occur at the most intense moment of meditation. The meditator experiences pure awareness, there is no discrete reality, no sense of passage of time, no sense of space, and there is no self-other dichotomy. This notion of Ultimate Reality will be further analysed and evaluated below.

does not give us any hint concerning the techniques that the meditators used it is rather difficult to determine which ways of experiencing could be relevant.

Persinger also considers spontaneous religious experiences, which may occur during personal crises and even during seizure experiences and Transcranial Magnetic Stimulation of the temporal lobes. (As mentioned earlier, these possibilities cannot be excluded). However, since Persinger does not regard the descriptions of the experiencers (except that he notes that they felt a sensed presence, vibration, or other experiences which he categorizes as complex partial epileptic like signs), it is difficult, if not impossible, to determine the ways in which the experiencers could have religious experiences.[89]

Furthermore, Persinger does not study these religious experiences. The experiences that he studies are epileptic seizure experiences. It is seizure experiences that are studied by way of neuroimaging technology and not religious ones, not those obtained through meditation or spontaneous ones. Neither does he take into account the neuroimaging studies performed on religious experiences by other neuroscientists.

For the same reason it is not possible to determine which types of spontaneous religious experiences were studied. Since there are no descriptions by the experiencers concerning what they experienced during their meditation sessions, I can only speculate concerning which types could be relevant.

4.8 Summary and Conclusions

The scientific validity of the neuropsychological studies of Persinger (1993b), (1993c), (1997) and (2003) were found to be questionable for the following reasons. Firstly, it was shown that the PPI that is used is clearly biased, because the questions (based on epileptic seizure experiences) he uses already contain the answers he wants to receive (that religious experiences are related to epileptic seizure experiences). Secondly, it was suggested that to make his study trustworthy, Persinger should have used at least one of the following tests: The Rey-Osterreith Test, which tests the non-dominant temporal lobe functions, and/or the Wechsler Logical Memory Test which is a verbal memory test for dominant lobe function. Thirdly, it was concluded that he could have used imaging technology like PET, SPECT or fMRI to study neural activity in the brain during religious experiences for the same reason. Fourthly, the sample he used for his investigation concerning the implications of religious experiences on the behav-

89 Any way mentioned above could be relevant.

iour of men and women is clearly not representative for the population of study, i.e., men and women having religious experiences. The sample only consisted of 50 right-handed male and 50 right-handed female persons. There was, for instance, no left-handed male or left-handed female control group. Besides this the sample did not consider persons younger than 18 or older than 25. It was also obvious that he did not consider people from different educational backgrounds (all participants were first-year undergraduate psychology students), and nor did he consider different social or religious backgrounds. Fifthly, it was questionable whether the TMS measured what it purported to measure, namely religious experiences of men and women. None of the items on the table of results is characteristic of religious experiences and the method was considered to cause seizure experiences when too high frequencies were applied. Furthermore, Granqvist et al., failed to replicate Persinger's results. They did not succeed in provoking any kind of such experiences by way of the TMS. One of the reasons was that Persinger did not use a double-blind study as was used by the Uppsala team. Finally, the scientific value of Persinger 1993b was shown to be questionable because the vectorial cerebral hemisphericity explanatory model, the PET-imaging as used here, and the TMS experiment do not measure, not even together, what they purport to measure.

It was also maintained that Persinger's statement, if we find what neural activity underlies epileptic seizures we have found what neural activity underlies religious experiences, is doubtful because it was shown that not all neuroscientists agree that there is a natural relationship between religious experiences and epileptic seizure experiences. Furthermore, religious experiences and epileptic seizure experiences have different neuroanatomies. To compare these experiences is similar to comparing a cold with hay fever. The symptoms of a cold are very similar to the symptoms of hay-fever; however a cold and hay-fever are completely different things. A cold is a virus while hay-fever is an allergy type. Their correlating neural activities are different. Another difference between religious experiences and epileptic seizure experiences is that the former are considered to be the result of communication between God or Ultimate Reality and the experiencer, while the latter might be seen as such by the epileptic person but are usually not.

Persinger's statement that right-handed men and left-handed women are potentially dangerous when having a religious experience because they may use these religious experiences, which they may interpret as hearing the voice of God, as justification for their aggression, and hence kill in the name of God, was shown to be based on speculative grounds. There was no significant neuroscientific support for Persinger's thesis that there is a specific correlation between hemisphere dominance and gender. Even though neuroscientists agree that there are biological differences in the male brain (the left-right differences being better articulated) and female brain (the front-back differences being more articulated), one cannot draw the conclusions Persinger draws.

Concerning the model of Vectorial Hemisphericity, it was maintained that neuroscientists seem to agree that the right hemisphere is more associated with non-verbal thought, visual spatial perceptions, and the perception, modulation and expression of emotions than the left hemisphere, and that the left cerebral hemisphere is more analytically inclined, and is more associated with language and mathematical processes than is the right hemisphere. Thus we could accept that RErl are associated with the right hemisphere (visions, hearing a voice) and RIE with the left hemisphere (understanding and interpretation). However, this does not mean that religious experiences can by default be provoked by TMS or that the neurological substrates that generate them can be isolated, copied and used for medical purpose in the sense that Persinger intends, because there are certain norms that a religious experience has to conform to. It has to be both recognised as such by the experiencer and by the religions (and continuities) through the doctrines and soteriology central to them, which is not what the TMS provokes.

We saw that Persinger concluded that experiences of Ultimate Reality are generated by the biology and/or psychology of personal crises and that religious experiences are illusory and that God or Ultimate Reality cannot exist. He explained religious experiences in terms of evolution, because according to him, such a biological capacity was necessary for the human species to survive. His conclusion was also that should the brain have developed differently, the God Experience would not have occurred. Once again Persinger maintained that religious experiences have nothing to do with God or Ultimate Reality. What happens, according to him, is that people re-experience their childhood relationship with their parents or parent-substitutes and thus, people simply mistake this experience for a religious one. Several possible supports for Persinger's idea were presented.

The first came from neuroscientists who maintain that when the corpus collosum of the brain of epileptic persons is cut, their seizure experiences (religious experiences to Persinger) disappear. This support was shown to be irrelevant because of the following reason. Even if an epileptic person had religious experiences which disappeared after the operation of the corpus collosum, it does not mean that the religious experiences the person had before the operation were illusory or that God could not exist.

The second derived from Churchland who asserted that Persinger's research on religious experience (Persinger 1993b and 1993c), drains probability from the hypothesis that the experiences provoking God-reports are truly experiences of God. However, if I am correct in my conclusion that we have good reasons to doubt the scientific value of Persinger's research on religious experience and that his research is biased, her support is also challenged.

Of course, it may be the case that people who have a religious experience during a personal crisis project the parental care or love of their infancy onto God and it may be the case that such a religious experience turns them into a more careful and loving person, i.e., they project God's care and love, as it were, onto their environment. It may even be the case that religious experiences have an evolutionary purpose. However, it cannot be concluded from this that religious experiences are illusory, or that there cannot exist any Ultimate Reality or God.

The religious experiences that Persinger was able to study are limited, even if he believes that his research covers the whole variety of religious and psychical experiences. Nevertheless, the religious experiences are similar to all other experiences in that the religious experiences obtained through different forms of meditation (mostly TM) are aspects of a subject and sometimes even have an object. Since Persinger did not take into account any description of what the experiencers experienced during their meditation, crisis, seizure, etc. and since he did not use any neuroimaging technology, it is not possible to determine which types of religious experience (as defined in chapter two) could be relevant for his studies.

Persinger's neuroscientific explanations of religious experiences are not able to explain what religious experiences are, which varieties of religious experiences exist, or whether religious experiences are real or illusory. Neither can they explain whether the brain creates or mediates religious experiences, nor whether or not God or Ultimate Reality exists. As they do not take into account the descriptions of the experiencers, Persinger's explanations cannot tell us what it is like to have such experiences, what consequences such an experience has for one's life, nor what is experienced. The problem with Persinger is that he explains religious experiences exhaustively in terms of something different such as epileptic like seizures, projected parental behaviour, an attribute of evolution for the survival of the human species, and in terms of nothing but, for example, abnormal temporal lobe activity or the shape of the neo cortex. Furthermore, religious experiences are illusory; the only thing that is real is what science explains religious experiences by.

5 Analysing Newberg and d'Aquili's Research

Since it is not possible to account for all their studies of religious experience within the framework of the current project, I have focused on those studies which I considered to be the most important for the purpose of the project. Hence, I shall analyse Newberg and d'Aquili's different neurological models for religious experiences obtained through meditation. The reason is that these models might be considered to have inspired Newberg and d'Aquili to undertake the neuroimaging experiment. It is important to note that there is no direct chronological order between the neuroimaging experiment and the other studies. The presentation of the experiment will also be different from the presentation of Newberg and d'Aquili's non-experimental research, because contrary to the experimental study, there are no explicit set-ups or hypotheses. However, my analysis will hopefully highlight the relationship between the different studies.

A short presentation of Newberg and d'Aquili and their work will facilitate our understanding of the intentions concerning their research on religious experience and hopefully the understanding of my interpretation of their work.

Newberg and d'Aquili's neuroscientific research is founded upon many years of research on neural activity that especially or in a special way could be involved in religious behaviour (for example, neurological structures that may underlie the establishment of myths and rituals and those involved in meditation).

In 1975, Eugene G. d'Aquili published his first article on the biopsychological determinants of religious ritual behaviour.[1] D'Aquili was a psychiatrist working in a private practice and a clinical associate professor of psychiatry at the University of Pennsylvania. He was also an anthropologist in addition to his medical occupation. Since the 1970s, his interest in understanding ritual and spiritual experience from a medical point of view led him to do intensive research on the matter, using all the methods at his disposal. The research received an experimental dimension after he met Andrew Newberg in 1991. Carol Albright says that d'Aquili

> was a visionary, a pioneer, and a stubborn soul, who followed his religious and intellectual calling along a path that was largely untravelled. He sought to explore the

1 d'Aquili/Laughlin (1975).

physical and spiritual bases of mystical experience and the implications of his findings. [...] In his pursuit of these understandings, d'Aquili seems to have been deeply influenced by the great mystics of Buddhism as well as by Catholic reflections from a variety of eras. Insights from these sources color his thought in significant ways.[2]

Thus, the fact that the neuroimaging experiment, which we shall consider below, involves Buddhist monks and Catholic nuns should not be unexpected.

Newberg was at that time still in medical school and worked at the University of Pennsylvania's nuclear-medicine department. Understandably, the neuroimaging work he was performing became an adjunct to d'Aquili's theoretical models. Today Newberg is the director of Clinical Nuclear Medicine and of NeuroPET Research, and associate Professor in the department of Radiology at the Hospital of the University of Pennsylvania. There, besides other tasks, he teaches the neurophysiology of religious experiences to the medical candidates, emphasizing the importance of spirituality. As in many countries, medical treatment has neglected this side of human nature. However, to Newberg it is important to understand the human being and the human world. To reach his goal he studied different religions as well as philosophy as a complement to neuroscience. For instance, he says that: "Religious experience involves a lot of different aspects of our thoughts, our emotions, our cognitions, and neuroimaging helps us look at what is happening."[3] However, he also asserts that "while there is great value in exploring the diverse variety of religious and spiritual experiences and practices using neuroscientific techniques, the religious and spiritual side must be carefully weighed, understood, and integrated into the analysis".[4]

Both Newberg and d'Aquili maintain that one has to take account of both the material and the spiritual side of the world. This is why they study religious experiences from a neuroscientific point of view bringing, as they put it, *neuro* and *theology* together. As mentioned earlier, their idea is to investigate, by neuroscientific means, how religious experiences can be understood from a neuroscientific point of view (neuro), taking the spiritual and other religious aspects into consideration (theology). It is important to note that they do not want to give religious experiences a neuroscientific explanation. Rather, they want to understand the correlation between experiences that are doctrinally and soteriologically central to religious traditions and what neuroscientific studies are able to show concerning these religious experiences. Nevertheless, they truly believe that neuroscientists may significantly contribute to the process of understanding religious experiences. Today neuroscientists are able to investigate what happens in the brain

2 Albright (2001) 485.
3 Kephart (2001) May/June, 3.
4 Newberg (2002) 146.

during certain religious experiences thanks to advanced imaging technologies such as PET, SPECT, fMRI and MEG.

I shall analyse and evaluate the following reasons which I believe guided Newberg and d'Aquili's research on religious experience. Firstly, they wanted to clarify which element(s), if any, is (are) invariant across the different religions. This is because they felt that previous explanations of religious experience had failed to do so, which according to them, may be the main reason why the soundness of religious experience is often dismissed. They believed that they could solve this problem and hoped that neuroscientific research on religious experience (and altered states of consciousness in general), would result in finding a typology of religious experience based on its correlating neural activity. Secondly, they wanted to find the specific correlation between religious experience and neural activity in brain functions and structures.[5] The experimental study may even have a third goal, namely, to allow Newberg and d'Aquili to confirm, confute or adjust their models for the correlation between the brain and religious experiences obtained through meditation.

The neuroscientific research of Newberg and d'Aquili which concerns my project therefore considers religious experiences obtained by way of different meditation techniques. They maintain for example that

> [f]rom a pragmatic point of view, [it] may be more interesting [to study] spontaneous mystical states rather than [...] those same states produced by various forms of controlled meditation. [...] But from the point of view of systematically developing a theoretical neuropsychological model of such states, analysis of controlled meditation gives us our best entree to an understanding of these phenomena.[6]

Thus one of the reasons why Newberg and d'Aquili chose to study religious experiences obtained by way of different meditation techniques is that studying such religious experiences allows them to understand the religious experiences both as obtained through meditation and as occurring spontaneously. Another reason is that spontaneous religious experiences are more complicated to use as the basis of neuroscientific studies. One of the problems is that it is not known which factors that trigger spontaneous experiences. This is not the case when the religious experience is obtained through meditation, since persistent training and endurance are known to be the factors that trigger them. For what concerns the experimental research, the experiments performed on religious experiences obtained through meditation can be repeated and tested in the way science prescribes, because the act of meditation can be performed in laboratory settings.

5 We are not simply talking about brain processes. It is important to distinguish between brain functions and brain structures in order to understand the neuroscientific research.
6 d'Aquili/Newberg (1993b) 186.

5.1 Neurological Models

Keeping in mind what is said above, let us consider their neurological models on religious experience obtained by way of different meditation techniques. Newberg and d'Aquili study the acts of meditation *via negativa* and *via positiva* meditation (among others). From what is said earlier we also know that the Catholic mystical tradition makes a distinction between acquired prayer, which is similar to via positiva meditation, and infused contemplation which was said to be similar to via negativa meditation. As mentioned in chapter two, the via negativa is a meditation technique in which the meditators concentrate on emptying their minds of all thought. The via positiva is a meditation technique in which the meditators focus on a chosen image. This image may be an inner image or an image situated somewhere in the room where the meditation takes place, for example, a cross on the wall or altar. On the basis of their knowledge of the functions of the brain, Newberg and d'Aquili outline hypothetical models for what the brain activity might look like during different acts of meditation. These models are hypothetical illustrations of which parts of the brain should have altered neural activity (compared to baseline neural activity) during the process of meditation. This process begins when the meditator begins to concentrate on emptying his or her mind of all thought using the via negativa (or when the meditator starts to focus on a chosen image using the via positiva), through the different phases in between to its final result, i.e., the experience at the most intense moment of meditation, which Newberg and d'Aquili called *Absolute Unitary Being*. In other words, different states of experience are linked to expected neural activity in the corresponding brain structure(s).

Newberg and d'Aquili also designed a scale on which they range different types of experience which they called the Aesthetic-Religious Continuum, because they believe that "to be successful, any model must be able to incorporate both relatively mild religious experiences and the most profound and compelling experiences".[7] This scale is, roughly, a continuum of experiences ranging from basic experiences (I am cold), through the realm of aesthetic experiences (watching the opera Don Giovanni) into the realm of religious ones (feeling God's presence in the forest, feeling religious ecstasy, Unio Mystica, etc.) and ending in experiencing Absolute Unitary Being (Ultimate Reality or God). With this scale, Newberg and d'Aquili want to account for human capacity to experience different conscious and emotional levels, and, for example, allowing oneself to be absorbed by a piece of music or to be in love. However, Newberg and d'Aquili do not put being absorbed by love on the same experiental level as being absorbed by

7 Newberg (2001b) 503.

music or art. As they see it, romantic love is a transitional phase between the aesthetic and religious experiences.[8] Without the capacity of being absorbed by something, no religious experiences would be possible according to them. The model of the Aesthetic-Religious Continuum is also valuable for Newberg and d'Aquili, since it enables them to relate the descriptions of the meditating and praying persons to specific brain functions and structures.

How are we to understand the experience which Newberg and d'Aquili regard as the most elevated religious experience possible, *Absolute Unitary Being*? There are two possible ways in which we can understand what they mean by Absolute Unitary Being. A first possibility is to regard Absolute Unitary Being as Ultimate Reality in the way I intended, namely, as the absolute ground of everything that is (and ever will be). Comprehended as such, *Being* means *to be*, *sein* in German, YAWH which means to be, to exist but also to become. Absolute Unitary Being is, then, equal to Absolute Unitary Sein.[9]

The second possibility is to understand Absolute Unitary Being in terms of *An* Ultimate Being. Thus, *Being* means *A Being*, *ein Wesen* in German. Absolute Unitary Being should then be understood as an Absolute Unitary Wesen.

Distinguishing between the two ways of understanding Absolute Unitary Being, i.e., being as sein and being as Wesen, is very important because this has consequences for how we should understand Newberg and d'Aquili's research, their assumptions and conclusions.

Since they do not intend to affiliate Absolute Unitary Being with a specific religion, (one of their aims is to find an invariant element across all religions), I believe that they understand Absolute Unitary Being as Absolute Unitary Sein and not as an Absolute Unitary Wesen. Unfortunately, as it will become clear, they do not always use the concept in a consistent manner.

Newberg and d'Aquili regard Absolute Unitary Being as *the* ultimate experience that can be achieved, because it is described as such by the meditators themselves. The meditators describe Absolute Unitary Being as not including any perception or awareness of discrete beings, nor of space or time. Furthermore, there is no self-other or self-world distinction experienced. It is also important to distinguish between the Absolute Unitary Being experience before it is put into words, the RErl and the interpreted experience, the RErf.[10] Thus when I refer to the Absolute Unitary Being

8 d'Aquili/Newberg (2000) 43. Unfortunately, it is not clear why Newberg and d'Aquili consider romantic love to be a link between aesthetic experiences and religious ones. Schopenhauer for instance, would regard altruistic love and not romantic love as the possible experiental bridge between aesthetic and religious experiences – Schopenhauer (1992) 524–527.

9 See further Runehov (2006).

10 I shall return to this, for now it will suffice to keep this distinction in mind in order to understand why I do not consider Absolute Unitary Being an experience *of* something.

experience as RErl I shall describe it as an *Absolute Unitary Being experience* and when I mean the experience as RErf I shall describe it as an *experience of Absolute Unitary Being.*

In order to facilitate an understanding of the neurological process of meditation, I have added a picture of the brain below, which illustrates the parts of the brain that showed altered neural activity during meditation. In addition, the process of via negativa and via positiva will be presented by way of charts. These will be presented in connection with the presentation of the meditation techniques respectively.

The Religious Brain

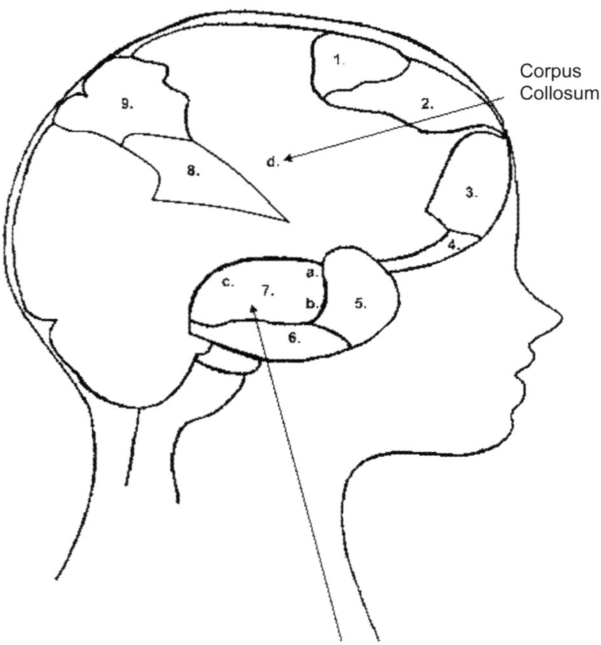

PREFRONTAL CORTEX (PC)
1. Supplementary Motor Area
2. and 3. Lateral Convexity of the Frontal Lobe
4. Orbital Frontal Cortex

TEMPORAL LOBES (PL)
5. Superior Temporal Lobe
6. Inferior Temporal Lobe (ITL)
7. Middle Temporal Lobe
8. Inferior Parietal Lobe (IPL)
9. Posterior Superior Parietal Lobe (PSPL)

LIMBIC SYSTEM
a. Hypothalamus
b. Amygdala
c. Hippocampus

Figure 1: The Religious Brain

5.1.1 The Via Negativa

The idea of the via negativa meditation technique is to silence all thought completely.[11] This type of meditation is also known as Monastic Mystical Consciousness, which implies that the mind has progressively emptied itself.[12] The meditator sits or lies down in a preferred position and focuses on emptying his or her consciousness of all thought by concentrating on, for example, his or her breathing. It is mostly (but not only) used in Eastern religions such as Buddhism and Hinduism, as well as in non-religious meditation. However, because the via negativa is mostly applied in Eastern religions this does not, of course, exclude religions other than Eastern ones involving via negativa in their tradition.

Newberg and d'Aquili hypothesize that the following neurophysiological processes should occur during via negativa meditation.[13] As the meditator concentrates on his or her breathing and thereby tries to silence all thought completely, impulses from the *prefrontal cortex* (PC) should inhibit the transmission of information from the right *inferior parietal lobe* (IPL) to the right *posterior superior parietal lobe* (PSPL).[14] That the meditation process seems to develop in the right parts of the brain is because Newberg and d'Aquili associate the right hemisphere with non-verbal awareness of the environment and the left hemisphere with speech, thought and analytic reasoning.[15] Their hypothesis is based on the following neurological knowledge: (1) The PSPL's neurological task is to analyse and integrate visual, auditory, olfactory and somatosensory information (see, hear, taste and touch) and to create three-dimensional pictures of the body in space. The PSPL is also responsible for the self-other and self-world distinction. (2) The prefrontal cortex (PC) is associated with our ability to focus upon important tasks and is called the seat of the will. Additionally, the PC plays an important part in memory, intelligence, concentration and higher cognitive functions including behaviour and emotions, i.e., our personality. Thanks to this part of the brain we can set goals, make plans and judge our priorities. In other words, the prefrontal cortex is the controller of intention. (3) The inferior parietal lobe (IPL) interprets sensory signals from other areas of the brain such as vision, hearing, motor sensory and memory and generates abstract concepts and conceptualizes them.

---

11 Via negativa actually means the path of negation.

12 Wainwright (1999) chap. 5.

13 d'Aquili/Newberg (1993b) 186–191/(1999) 110–114.

14 It might be illustrative to have figures 2 and 3 at hand, because it shows what happens in the brain in an educative way using the same terminology as the full account.

15 d'Aquili/Newberg (1993b) 178–179.

This then means that when the meditator focuses on silencing all thought, the prefrontal cortex (PC) which is the seat of the will, should inhibit transmission of information from the right inferior parietal lobe (IPL), which is the interpreter of different visual, auditory and other signals, to the right posterior superior parietal lobe (PSPL). This brain area is supposed to analyse and integrate these visual, auditory and other signals that the IPL has interpreted.

When these sensory signals are systematically blocked by the meditator's will to silence all thought, partial deafferentation should occur in the right PSPL, because, due to the meditator's blocking act, the right PSPL should receive less and less information to analyse and integrate. In other words, information should be prevented from reaching the part of the brain which normally processes them.

As the meditator continues to focus on emptying her or his consciousness of all thought, the right prefrontal cortex should stimulate the right *hippocampus*, which is located slightly behind the *amygdala* in the *temporal lobe*. The temporal lobe is situated at the sides of the head and is associated with language and conceptual thinking. These two often act in a complementary way to focus the mind's attention on interesting sensory input, to generate emotions, and to link those emotions to images, memory, and learning. The hippocampus is associated with the creation, storing, processing and encoding of memories and finding memories, as well as with preventing emotional extremes. The hippocampus is often called the *diplomat*. The meditator is still concentrated on his or her breathing which implies that the hippocampus in its turn, should stimulate the right amygdala *trophotropic centre*. The trophotropic centre is part of the parasympathethic nervous system and is associated with quiescence. For example it slows down the heart rate.[16] The amygdala should then stimulate the *hypothalamus*, which is often called the master controller. The amygdala, which is located in the middle part of the temporal lobe, is one of the oldest structures in the brain and controls and mediates virtually all high-order emotional functions. It is maintained that if you become aware of a suspicious noise in the middle of the night, it is not your cat or dog that has awakened you but your amygdala. This is because the amygdala is responsible for quickening your pulse and triggering the surge of fear that makes your eyes snap open. Another example of the amygdala's alertness is when you meet the *woman* or *man* of your life. In that case your amygdala will see to it that you respond in the

16 Its counterpart is the ergotropic centre which is part of the sympathetic nervous system and which is associated with arousal. It will accelerate the heart rate. The parasympathetic and the sympathetic nervous systems are two sides of the autonomic peripheral nervous system or ANS which is a part of the peripheral nervous system (the other part is called the somatic peripheral nervous system) and is connected to the central nervous system or CNS. The ANS serves various organic structures such as the heart, the blood vessels, the digestive systems, the genital organs and so on.

correct emotional way. Therefore, the Amygdala is often called the *watch-dog*. Like the amygdala, the hypothalamus belongs to the oldest structures of the limbic system. It is crucial to our survival because it regulates the arousal and quiescent systems. In other words the "fight or flight" response as well as the arousal of fear and positive emotions ranging from moderate pleasure to bliss.[17]

At this point in the via negativa meditation process, the meditator feels relaxed and his or her mind has become increasingly free of thoughts. The temperature of the body has decreased and breathing has become slower. Nevertheless, Absolute Unitary Being is not yet obtained. In Newberg and d'Aquili's view there is not even a religious experience. This has to do with their Aesthetic-Religious Continuum model mentioned above. For Absolute Unitary Being to be obtained the following neurological processes have yet to occur.

When the meditator continues to focus on emptying her or his consciousness of all thought, the right PC stimulates the right hippocampus, the memory-manager, which in its own turn stimulates the right limbic system (the right amygdala's trophotropic centre and the right hypothalamus), the seat of emotions. The meditator will now feel relaxed.

While the meditator continues to focus on emptying his or her consciousness of all thought, the right hypothalamus should return neural signals to the right PSPL by way of the right amygdala and the right hippocampus. Again the right PSPL should stimulate the right hippocampus in a loop-like process which will repeat itself until a spillover occurs in the right hypothalamus. This spillover should cause a maximal stimulation of the *ergotropic sympathetic system*, which will finally cause a spillover even there. Newberg and d'Aquili suggest that because the ergotropic sympathetic system is associated with arousal, the meditator will now obtain an ecstatic and blissful feeling but not yet Absolute Unitary Being.

Newberg and d'Aquili suggest that after this state of meditation, while the meditator continues to focus, the neurological processes of via negativa meditation should proceed as follows. The hypothalamus should simultaneously and instantaneously stimulate the left and right amygdala, the left and right hippocampus and the left and right prefrontal cortex (PC). Finally, when the PC is maximally stimulated, a total deafferentation of the left and right posterior superior parietal lobe (PSPL) should occur and it is this deafferentation that allows the meditator to experience Absolute Unitary Being, since the PSPL is meant to maintain the self-other and self-world relation and since it is totally blocked (deafferented), by the meditator's

17 It also regulates many hormonal systems including reproductive hormones, the thyroid and growth hormone, and also moderates the immune system, hunger, thirst, and body temperature.

concentration.[18] Newberg and d'Aquili thus maintain that what the meditators describe as their ultimate experience, an absolute homogeneous unlimited union, correlates with a total deafferentation of the left and right PSPL. The PSPL is often referred to as *the self.*

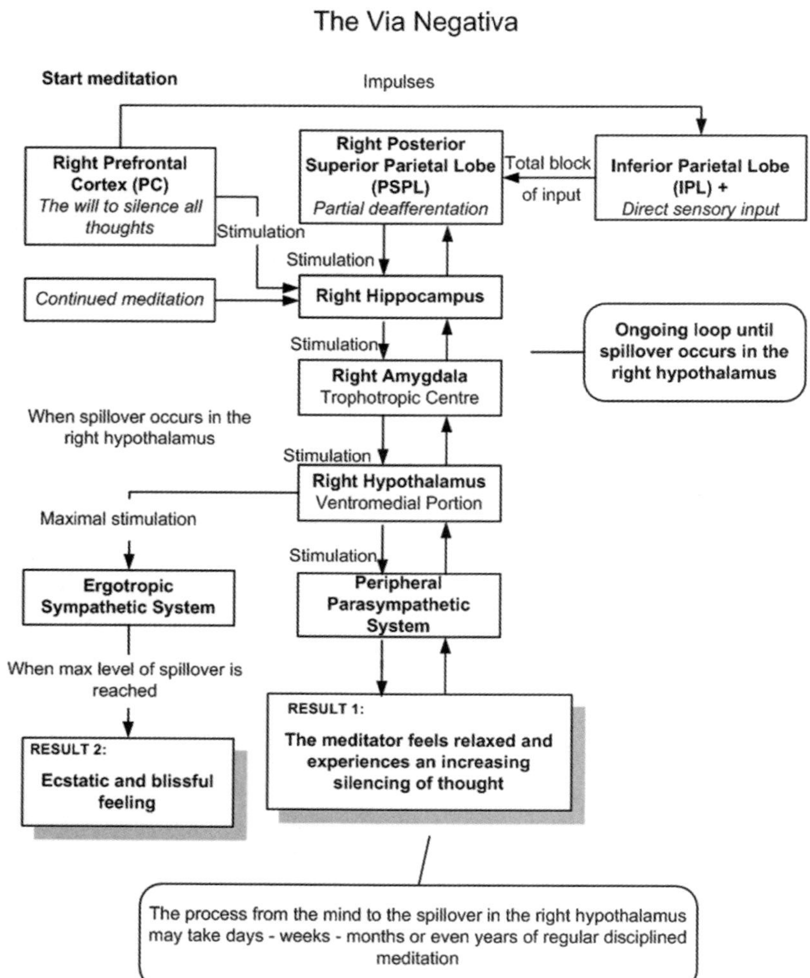

Figure 2: The Via Negativa: first stage

---

18 Deafferentation means that all neural input or flow into a certain brain structure is cut off. This cutting off may be partial or total. It is the latter that is associated with Absolute Unitary Being. The total deafferentation of both the left and right posterior superior parietal lobes is considered as a *final* point of meditation neurologically. Thus there seems to be a conformity between the narratives of meditators and neuroscientific theories.

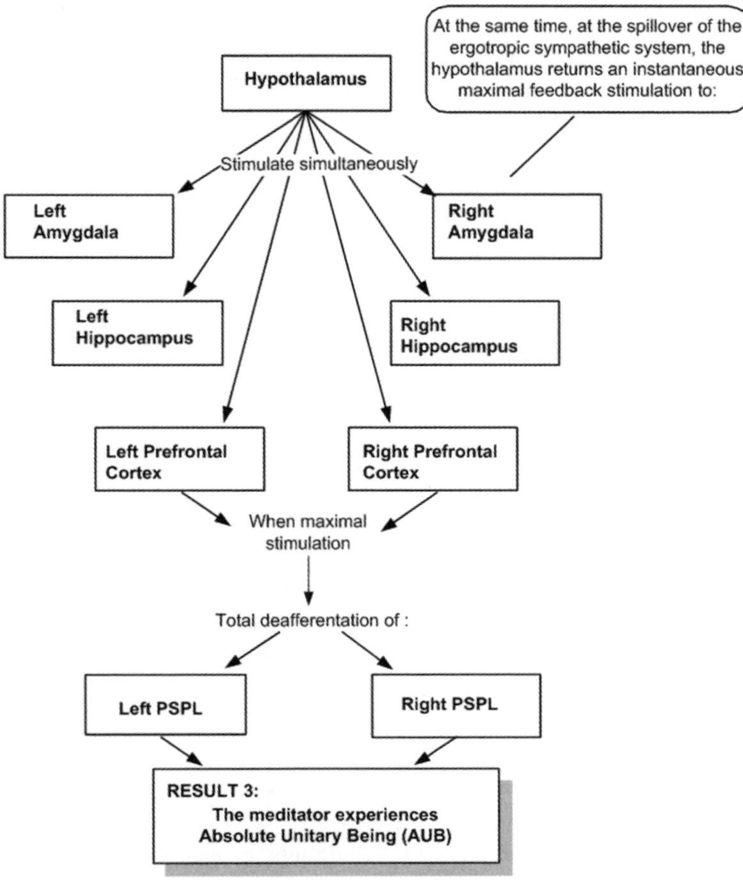

Figure 3: The Via Negativa: final stage

What then do Newberg and d'Aquili's neurological models for religious experiences obtained by the active or via positiva meditation look like? Again a flowchart of the model is presented in direct connection to the presentation below.

5.1.2 The Via Positiva

The idea of the via positiva meditation technique, then, is instead of focusing on emptying his or her consciousness of all thought, the meditator focuses on an image. Christian mystics, for example, often use this meditation technique.[19] Newberg and d'Aquili differentiate between active meditation

19 See also Wainwright (1999), chap. 5.

where the meditator focuses on a visual preferred object in the external world (with open eyes), and active meditation where the meditator focuses on a preferred image from memory (with eyes closed). The difference is, obviously, that the visual system is more involved in the former method.

Newberg and d'Aquili hypothesize that the following neurophysiological processes should occur during via positiva meditation.[20] Similar to the via negativa, the process should begin by how the right prefrontal cortex (PC) responds to the meditator's concentration. Then, the prefrontal cortex should send neural impulses to the right *inferior temporal lobe* (ITL) while simultaneously stimulating the right PSPL to fix the image. For our purpose it is important to know that the ITL detects objects from the PSPL and mediates the objects of interest and importance. The right ITL should then stimulate the lateral portions of the right hypothalamus via the right hippocampus and the right amygdala. The result should be, Newberg and d'Aquili say, that the meditator feels a pleasant sensation.

When the meditator focuses on an image, say a blue lagoon, neurologically the PC sends neural impulses to the right inferior temporal lobe (ITL) which mediates important images and, at the same time, stimulates the right PSPL (the analyser of visual, auditory and other information) to fix the image (the blue lagoon). Then, the ITL will stimulate the right hippocampus which stimulates the right amygdala which will stimulate the right hypothalamus. Now the meditator should feel a pleasant sensation.

Similar to the via negativa, a loop-like process should be established which increases in intensity until the stimulation of the lateral hypothalamus, more precisely the ergotropic system, has maximal stimulation. A maximal stimulation of the right hypothalamus, especially the trophotropic system (associated with quiescence) occurs and causes a spillover. Simultaneously there should be an ongoing maximal stimulation of the ergotropic system (associated with arousal). Because of that the whole limbic system is stimulated to the maximum, which results in a total deafferentation of the *left* PSPL via maximal stimulation of the left and right prefrontal cortices. Observe that the *right* PSPL is only partially deafferented. The meditator, Newberg and d'Aquili write, feels ecstatic.

Now, Newberg and d'Aquili suggest that there is a period of conflict on the neural level between the self-other dichotomy that has been eliminated (caused by the total deafferentation of the left PSPL) and the meditator's persistent focusing on the image (resulting in a partial deafferentation of the right PSPL). Meditators remember this state as the time when he or she

---

20 d'Aquili/Newberg (1993b) 191–194/(1999) 114–116. It might be illustrative to have figures 4 and 5 at hand, because it shows what happens in the brain in an educative way using the same terminology as the full account.

feels absorbed into the object of meditation (becomes one with it), here the blue lagoon. The length of the union depends on how intensely the meditator focuses on the object of meditation. Hence, a union with the object of meditation or prayer is, neurologically, associated with a neural conflict in the PSPL.

According to Newberg and d'Aquili, two different situations may occur: (1) Reinforcement of the experience of that union by continuous focusing on the object and thus stimulating the ergotropic structures or (2) a switch to a profound quiescence by allowing the trophotropic structures to take over. In the latter situation, the meditator stops focusing on the image which results in the deafferentation of the right PSPL. As the left PSPL is already deafferented, the meditator will experience Absolute Unitary Being. A good example of how the focusing on an image turns into an Absolute Unitary Being experience is:

> But I perceive, not with my flesh eyes, which look on this icon of Thee, but with the eyes of my mind and understanding, the invisible truth of Thy face, which therein is signified, under a shadow and limitation. Thy true face is freed from any limitation, it hath neither quantity nor quality, nor is it of time or place, for it is the Absolute Form, the Face of faces.[21]

The religious experiences obtained by way of via negativa and via positiva meditation techniques are of course not the only possible ways to undergo religious experiences. Newberg and d'Aquili are well aware of the variety of religious experiences and have no intention, they maintain, of creating a model for all of them, or a model including all of them. Yet they have designed a neurological model for what Rudolf Otto called *numinosum* and *mysterium tremendum et fascinosum*.[22]

---

21 Nicholas of Cusa, The Vision of God, cited by Peterson et al. (2003) 19.

22 Otto (1971), d'Aquili/Newberg (1993b) 195–196/(1999) 102–106. I shall not consider this theoretical model, because I do not believe that it supplies any further information that would be of interest for my purpose.

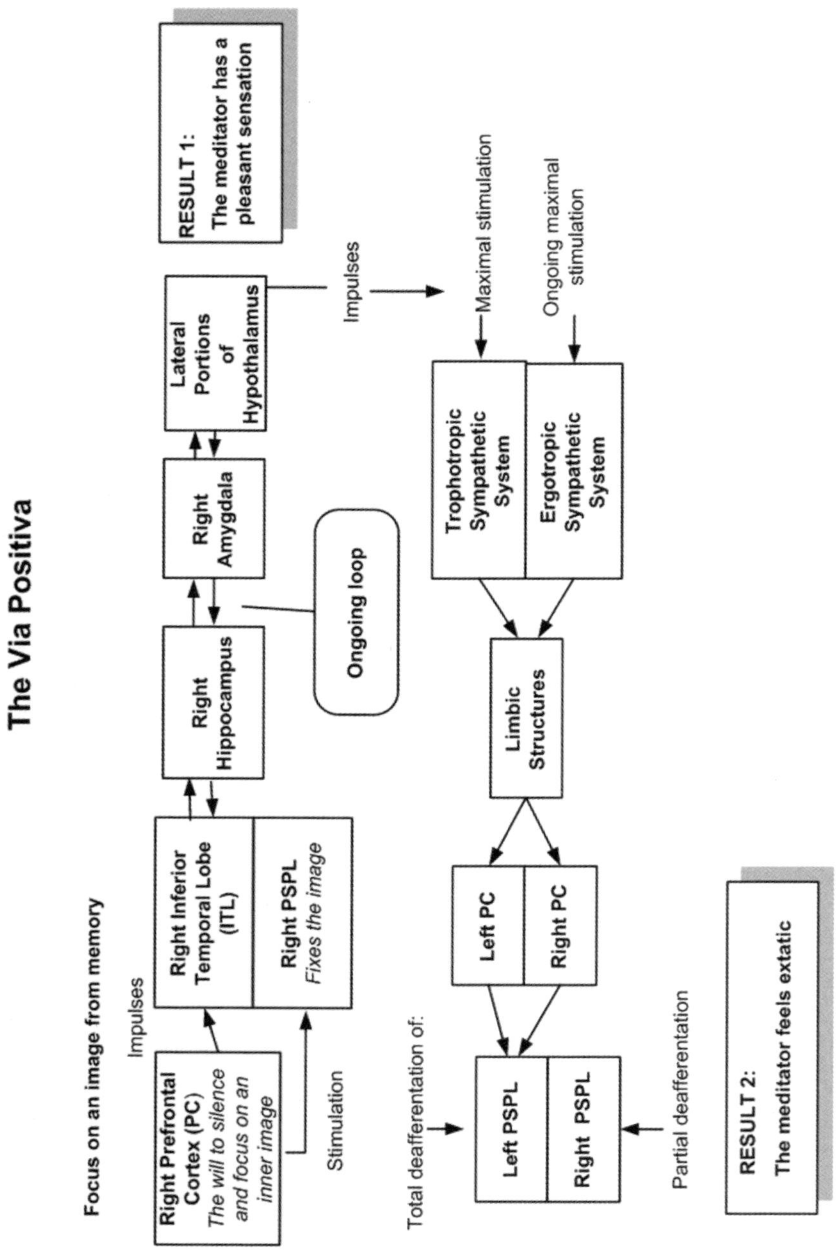

Figure 4: The Via Positiva: first stage

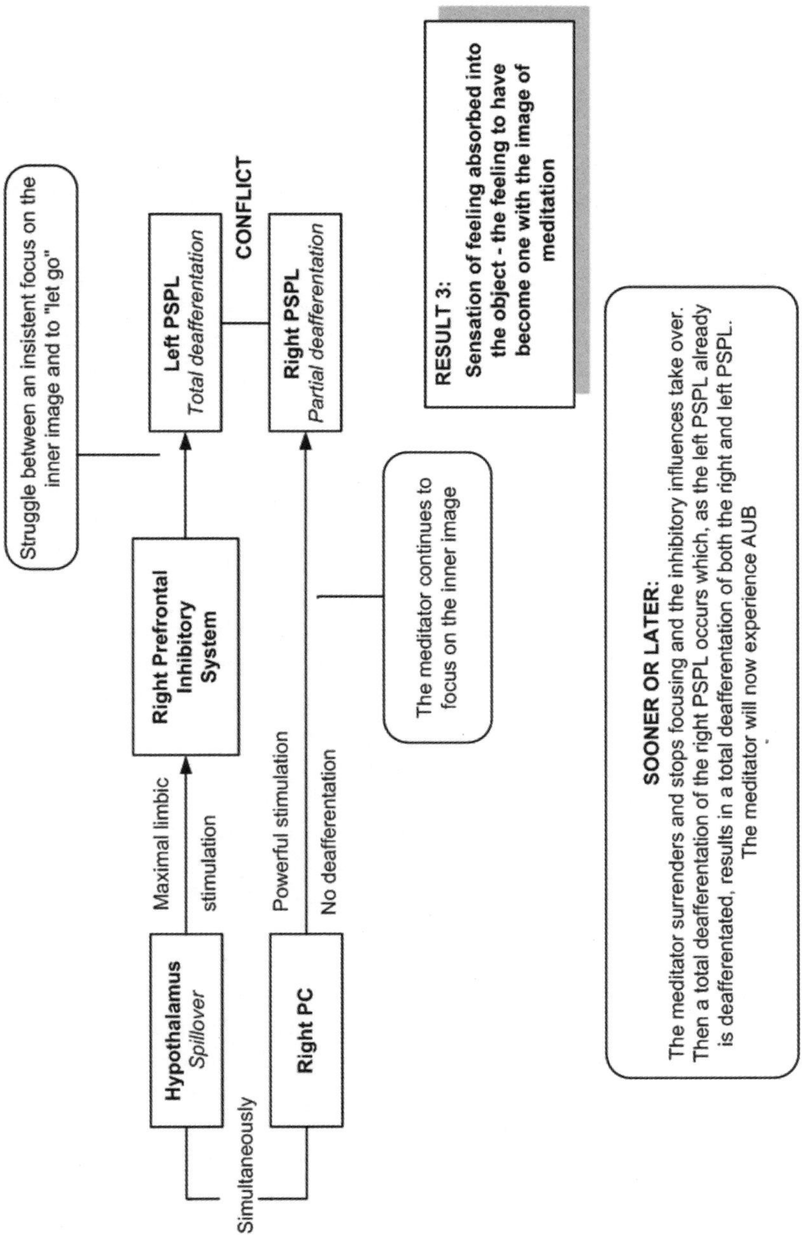

Figure 5: The Via Positiva: final stage

Andrew Newberg and Eugene d'Aquili thus had an idea of how different religious experiences that could be ranged on the Aesthetic-Religious Continuum should correlate with neural activity in different brain structures. They also had an idea about which differences would distinguish the via negativa from the via positiva. Finally, they suggested that the ultimate experience, Absolute Unitary Being, should correlate with a total deafferentation of both the left and right posterior superior parietal lobes. They write:

> It is certainly likely that there are parts of this model which will be proved incorrect, but it is our hope that many of the basic aspects of the model are indeed correct. Furthermore, the ability in itself to describe mystical states from a neuropsychological perspective, whether the model proves accurate or not in all aspects, is an important step in understanding how the human brain relates to religious experience.[23]

5.2 Experimental Research

Thus, it seems that they had in mind that their results would be tested by way of a neuroscientific experiment, whether or not they would perform it themselves, in the hope that their neurological models for religious experiences would be proved accurate. Hence, one of their reasons for carrying out an experimental study of religious experience was probably that they wanted to see whether the previous hypotheses of which structures of the brain should correlate with different types of religious experience could be acknowledged. Could these hypotheses be confirmed?

Newberg and d'Aquili chose to study different acts of meditation and prayer, because firstly, experiences obtained through meditation or prayer, unlike other forms of religious experiences, are repeatable, because practitioners of meditation and prayer are able to actualise their experiences in laboratory settings. Secondly, meditation and prayer are complex neurocognitive tasks often associated with physiological and psychological changes which are measurable by neuroscientific means. Meditation and prayer activate the brain in specific ways and it is this brain activity that can be measured. Meditation and prayer are neurocognitive tasks. The experiment was performed as follows.[24]

Set Up and Method

The initial subjects of the experiment were eight experienced Tibetan Buddhist meditators. Newberg and d'Aquili later broadened their experiment by

23 d'Aquili/Newberg (1993b) 196.
24 Newberg et al. (2001a).

adding several Franciscan nuns at prayer. All these persons had more than 15 years of practice in meditation or prayer. They practised at least one hour per day and at least five days a week. Furthermore, all of them had participated in several retreats lasting a minimum of three-months and in one annual retreat of one month.[25] The Tibetan Buddhists were four women and four men, between the ages of 38 and 52 (the average age being 45). None of the subjects had a medical or (neuro)psychological history nor did they have any history of drug abuse. There was also a control group, existing of nine healthy persons who did not perform any act of meditation but who also underwent a SPECT-scan at baseline. The experiment took place in a small dark room at the laboratory of a large university hospital. All were tested separately. Let us follow one of the meditators called Robert.

Robert sat down on the floor in the lotus position, surrounded by candles and sticks of jasmine incense. He was injected at baseline (start position) with 7 mCi of a specific radioactive tracer called HexaMethyl Propylene Amine Oxime (HMPAO) and brain-scanned 20 minutes later in a Single Photon Emission Computed Tomography (SPECT) during 45 minutes.[26] In this way, pictures were taken of his brain in its basic state, i.e., the state before beginning the act of meditation. Newberg and d'Aquili registered the values for the regions of interest in the brain. Robert began his meditation. He was connected to the research team by way of a common cotton twine which he had to pull as soon as he felt that he approached his most intense moment of meditation. When he had been meditating for about an hour he pulled the twine and was injected with 25 mCi HMPAO. Twenty minutes later Robert was SPECT-scanned during 30 minutes. Newberg and d'Aquili registered the new values. All the subjects, both the Tibetan monks and the Franciscan nuns, were examined in exactly the same way.[27]

Knowing that there are several advanced instruments used for image-diagnostics in the medical world today, the question arises why Newberg and d'Aquili chose to use the SPECT-camera. It seems that their choice was limited. Newberg and d'Aquili had to choose to use SPECT as an imaging technique in preference to functional Magnetic Resonance Imaging (fMRI) and Positron Emission Tomography (PET), because even though the fMRI has improved resolution as well as the ability to provide immediate anatomic correlation compared to SPECT, it makes too much noise. As a matter of fact, Newberg and d'Aquili tried the fMRI first on a subject, but the meditator found it impossible to carry out the meditation. Even the PET

25 Newberg et al. (2001a) 115. The name Robert is taken from Newberg et al. (2001b).
26 mCi means milli-Curie.
27 Newberg et al. (2001a) 115.

provides better resolution than SPECT. However, Newberg and d'Aquili wanted to make the environment as free of distraction as possible. But they do not explain why the environment is better in this aspect when using a SPECT camera instead of a PET camera.[28]

They present the following hypotheses, which they drew from previous neuroscientific studies of religious experience done by themselves as well as by others. Their hypotheses prior to their experiments were:

Hypotheses

1. Because earlier Electroencephalogram (EEG) studies of meditation as well as studies of attention-focusing tasks had resulted in frontal activity, they expected the SPECT images to show *increased activity in the frontal lobes, in particular the prefrontal cortex during meditation.*[29]
2. Because meditation is associated with alterations in the subjective experience of space and time and the posterior superior parietal lobe has to do with visual-spatial and temporal processing in the brain as well as body orientation, they also expected the SPECT images to show *decreased activity in the superior parietal lobe.*[30]

---

28 SPECT takes a series of images of the brain that will show how well blood flows through its various areas. The only noise is the mechanical rotation of the camera, which is only a slight noise. SPECT scanning is based on the principle of blood flow, because the areas in the brain with increased blood flow take up more radioactive tracer (for example HMPAO) than those with less blood flow. In other words, blood flow in the brain is directly related to brain activity. The most important difference between SPECT and PET is that SPECT allows you to scan the person up till 24 hours after the injection of the radioactive tracer. PET also has a slower uptake value than SPECT. Therefore, SPECT-imaging is used for investigating epileptic seizures. It is not possible to scan a person during his or her seizure but it is possible to inject the radioactive tracer. The person will not suffer from such interference. Within the following 24 hours, neurologists can investigate brain activity during a seizure by way of SPECT-imaging a person. Another difference between SPECT and PET is that PET imaging applies two positrons which are directed 180 degrees away from each other at the moment of disintegration, while SPECT only uses one single gamma ray. The latter implies that it is more difficult to determine the exact position and results in lower resolution. This might be the reason why tracers with a lower half-life are used such as HMPAO. Half-life also implies that the isotopes do not have to be produced on site, which means lower costs since the hospital/research institute does not need a cyclotron for isotope production. We should also note that PET, SPECT and fMRI are not the only neuroimaging facilities neuroscientists have at their disposal today, for example, there is also the *Magnetoencephalography* (MEG) which is 4-dimensional, and a direct measure of brain function, unlike functional measures such as fMRI, PET and SPECT that are secondary measures of brain function reflecting brain metabolism. Finally, sometimes spin-labeling is used. However, there is no need to enter further into the different neuroimaging techniques for our purpose.

29 The frontal lobes are known as the "brain's command post" according to Goldberg. Goldberg (2001); Newberg et al. (2001a).

30 Newberg et al. (2001a).

3. Because meditation is said to bring about a decrease in sensory awareness and motor activity, they expected the SPECT images to show *decreased activity in the sensorimotor area.*[31]
4. Because of autonomic changes associated with meditation, they suggested that this has to imply *changes in the midbrain.* [32]
5. As there are no previous results that show that the cerebellum, the superior frontal cortex and the occipital lobes are involved in meditation, they expected little change in these areas of the brain.[33]
6. Finally, Newberg and d'Aquili considered that there might be *baseline differences in reginal cerebral blood flow (rCBF) between the meditators and the control group*, because earlier studies suggested this.[34]

Results and Discussion

Generally considered, Newberg and d'Aquili observed that during the act of meditation, the meditators had increased rCBF (reginal Cerebral Blood Flow) in the inferior and orbital frontal cortices, the dorsolateral prefrontal cortices (i.e., the cortices situated slightly at the back of the side of the brain), the sensorimotor cortices, the dorsomedial cortices (i.e., the cortices at the back of the middle of the brain), the midbrain, the cingulate gyri and the thalami. The other parts of their brains did not show any significant changes in neural activity.[35]

More specifically, the result corresponded to the hypotheses in the following way. When the values of the baseline scans were compared with the values of the scans taken during meditation, they discovered that there was significantly increased activity in the frontal lobes, which implies that the first hypothesis was confirmed as well as previous studies of meditation by way of the EEG.[36]

However Newberg and d'Aquili *did not find* a significant decrease in the superior posterior parietal lobes as the second hypothesis suggested but only an overall *tendency* in that direction. The SPECT images taken of Robert's brain before meditation showed that there was furious neurological activity in this area of the brain. This activity appeared on the computer screen of the SPECT as vibrant bursts of brilliant reds and yellows. The images taken during the act of meditation however, showed the same area

31 Newberg et al. (2001a).
32 Newberg et al. (2001a).
33 Newberg et al. (2001a).
34 Newberg et al. (2001a) 115.
35 Newberg et al. (2001a) 117–122.
36 Newberg et al. (2001a) 117–122.

in green and blue, which are colours that indicate reduction in activity.[37] Remember that the darker the colours are, the greater the decrease in activity. Hence a total deafferentation would correlate with very dark colours on the computer screen of the SPECT.

The third hypothesis that there would be a decrease in activity of the sensorimotor areas during meditation was not confirmed: rather, the opposite occurred and an increase was measured. Newberg and d'Aquili guessed that in order to maintain the posture of the meditator there might be a significant degree of activity required in the sensorimotor areas. Furthermore, there may be some form of visual input from the internal images generated during this type of meditation. This would explain the unexpected increase of activity in the sensorimotor areas. However, they realised that the relationship between the sensorimotor cortex and meditative practices has to be studied more profoundly to be able to confirm or confute these assumptions.[38]

The fourth hypothesis that there have to be changes in the midbrain because of autonomic changes during meditation, was confirmed, as they found increased activity in the midbrain. Nevertheless, even here more experimental studies are needed, because they did not specifically measure autonomic functions, which, according to them, they should have done since changes in these autonomic nervous system functions are frequently associated with meditation. Meditators often report a decrease of blood pressure and heart rate.[39]

The fifth hypothesis predicted that there would not be significant changes in the cerebellum, superior frontal, or occipital areas, since the function of these parts of the brain is not normally associated with meditation, and this was confirmed. (The occipital lobe is concerned with vision.) However, and unexpectedly, they found significantly increased activity in the thalamus, something that may be important for the overall complex processes associated with meditation, including both cognitive and affective responses.[40]

Finally, the sixth hypothesis concerning the differences assumed in the brains at rest in the control group compared to the brains at rest in the meditators, Newberg and d'Aquili discovered that the meditators had a significantly different thalamic laterality index at baseline (before the act of meditation began) compared with the control subjects. Consequently, the differences in the thalamus led them to assume that the meditators might have undergone changes in their brain, because of their intense meditation practice. However, even though this is a very interesting finding, they do not

37 Newberg et al. (2001a) 117–122.
38 Newberg et al. (2001a) 117–122.
39 Newberg et al. (2001a) 117–122.
40 Newberg et al. (2001a) 117–122.

clarify it any further but maintain that they need to do more research to be able to understand and confirm or disconfirm this discovery.[41]

One important question which Newberg and d'Aquili did not ask, but in my view, should have asked, is whether they were able to detect any difference in neural activity between the persons practising different meditation techniques. The experimental neuroscientific study, as a matter of fact, showed increased neural activity in the sensorimotor areas, which indicated that some meditating or praying subject used a meditation technique that involved the visual structures of the brain. (For example, the via positiva meditation technique). Therefore, it is unfortunate that they did not include the hypothesis that a different neural activity should be expected depending on which meditation techniques the individuals applied. Now we do not know whether the result comes from Tibetan monks or Franciscan nuns or both. As this is an important issue, we shall return to it later.

In conclusion, firstly, the SPECT images which illustrate the parts of the brain which have increased or decreased activity during the meditation act in specific colours on the computer screen of the SPECT; secondly, the description is given by the meditators of what they experience during the act of meditation and, finally, in their previous neuroscientific work, Newberg and d'Aquili were able to complete their neurological models for some religious experiences. However, their experimental study could not entirely confirm their neurological models for the correlation between the brain and religious experience obtained by way of different meditation techniques.

5.3 Newberg and d'Aquili's Conclusions

The question now at stake is which conclusions Newberg and d'Aquili have drawn from their research and how their account of religious experiences should be understood. The following conclusions are at stake, (1) Absolute Unitary Being is the invariant element across all religions, (2) all experiences are equal, (3) Absolute Unitary Being or God really exists and (4) mind and brain are complementary.

5.3.1 The Invariant Element

As mentioned earlier, the meditators who have experienced Absolute Unitary Being describe their experience similarly in the sense that there are no discrete objects or beings, that there is no sense of space or time and that

41 Newberg et al. (2001a) 117–122.

there is no border between the self and the rest of the world. As a matter of fact, some say that in such a state, one does not experience the subject of self at all. There are no thoughts, no words and there is no sensation. Furthermore, according to Newberg and d'Aquili, there cannot be two different Absolute Unitary Being experiences as such, because there are "no competing versions of truth; there is only truth itself, so conflicting beliefs [...] are not even possible".[42]

That we may become confronted with different versions of Absolute Unitary Being is because when the experiencer mediates his or her Absolute Unitary Being experience, he or she puts it in the framework of his or her own culture, religion, and personal belief. Some will say that they felt united with God, others will say that they felt as a drop of water in the ocean, others will interpret their experience as an experience of total emptiness, nirvana, etc. In other words, the differences are due to the after-the-fact-subjectivity. In *Why God Won't Go Away*, Newberg and d'Aquili write:

> While in the state of Absolute Unitary Being, subjective observations are impossible; on the one hand, no subjective self exists to make them, and on the other, there is nothing distinct to be observed; the observer and the observation are one and the same, there are no degrees of difference, there is no *this* and no *that*, as the mystics would say. There is only absolute unity, and there cannot be two versions of any unity that is absolute.[43]

What Newberg and d'Aquili mean is that the Absolute Unitary Being experience at the very moment of the experience and before the experiencer puts the experience into words (i.e., as she or he remembers the experience later on) is equal for all. I believe they reasoned as follows.

From their models for the correlation between the brain and religious experiences they borrow the hypothesis that the Absolute Unitary Being experience (AUB) correlates to the total deafferentation of both the left and right posterior superior parietal lobes (PSPL). They were not able to confirm this hypothesis experimentally, but accounted for a tendency towards it. However, they say there cannot be two different total deafferented PSPL, because neuroimaging of various deafferented brain-structures shows exactly the same neural activity (here non-activity) in the PSPL. They conclude that there cannot be two different Absolute Unitary Being experiences. Hence

1. If there is an Absolute Unitary Being experience then there is also a total deafferentation of the left and right PSPL.

---

42 Newberg et al. (2001b) 163. Let us not concentrate on their use of the word truth, which is philosophically seen problematic, since this would take us away from my purpose.
43 Newberg et al. (2001b) 123.

2. Two different total deafferentations of PSPL are not possible
3. *Therefore*, there cannot be two different Absolute Unitary Being experiences

Because all meditators, regardless of which religion they belong to, (sooner or later) experience Absolute Unitary Being as their most intense moment of meditation, Newberg and d'Aquili believe that Absolute Unitary Being could be the invariant element across all religions, the mutual core for all religions. This suggestion is problematic, not least because not all mystics would agree that all Ultimacy Experiences are equal. Furthermore, as we also shall see in the evaluation chapter below, some neuroscientists maintain that a total deafferentation of the left and right PSPL is simply not possible.

However, the fact that Newberg and d'Aquili suggest that there cannot be two different Absolute Unitary Being experiences may also have to do with their second conclusion I want to discuss, namely, that all experiences are equal.

5.3.2 All Experiences are Equal

In *Why God Won't Go Away*, Newberg and d'Aquili write:

> All the things the mind perceives – all thoughts, feelings, hunches, memories, insights, desires, and revelations – have been assembled piece by piece by the processing powers of the brain from the swirl of neural blips, sensory perceptions, and scattered cognitions dwelling in its structures and neural pathways.[44]

Some neuroscientists would take the fact that religious experiences are associated with *observable* neurological activity as support for the hypothesis that religious experiences are nothing but brain activity, that God is merely *a ghost* in the brain.[45] These neuroscientists would argue that Newberg and d'Aquili's research on the relation between the brain and religious experiences *prove* that people's experience of God can be reduced to neurochemical flux. Newberg and d'Aquili write, however, that doing so

> [m]ay be equivalent to maintaining that their experience of the [...] sun, the earth, and the air they breathe is reducible to neurochemical flux. [And that] [a]ny attempt to trivialize their ontological status by naive reductionism might be a serious misinterpretation of the true nature of these mystical experiences.[46]

---

44 Newberg et al. (2001b) 36.

45 The brain activity shown on the computer screen of the SPECT camera of the brain of a person absorbed in the act of meditation.

46 d'Aquili/Newberg (1999) 120.

Hence, Newberg and d'Aquili treat all human experiences as equal from a neuroscientific point of view in the sense that every experience is associated with neural activity. To them, all experiences are real or all experiences are neurochemical flux. Thus, if we consider religious experiences to be nothing but neuron-activity, we have to consider all our experiences as such. Their idea is that seen under a SPECT brain scan, all the neurological activity involved in eating apple pie, the olfactory areas registering the aroma, the visual areas picturing the pie, the centres of touch perceiving the pie's surface and the areas responsible for taste, are shown in the same way as the different brain activities involved in religious experiences, i.e., as increased or decreased neural activity in the relevant parts of the brain. Does this imply that the experience of eating apple pie is not real? That it is not delicious? Not to Newberg and d'Aquili. It is their view that the experience of eating pie is equally as real, seen from a neuroscientific perspective, as experiencing Ultimate Reality or God.[47]

Thus, neurologically, the SPECT-images resulting from their neuroscientific experimental study of religious experience could be images of a person performing a cognitive-task, of which meditation is an example. The reason the study concerns religious experiences is because the subjects of the experiment were Buddhist monks and Franciscan nuns who frequently practise meditation and prayer and because their task in the experiment was to meditate and pray, and not some other cognitive exercise.

Similarly, if a person was brain scanned while eating a piece of apple pie, the SPECT-images resulting from this imaging examination could be images of a person performing a perceptive task, of which eating is an example. The reason this study, then, concerns experiences of eating a slice of apple pie is because the subject of the experiment very much liked apple pie and because their task in the experiment was to eat apple pie and not to perform some other perceptive task.

This will also be revisited in chapter six. However, it seems that Newberg and d'Aquili want to say something more than this.

5.3.3 Absolute Unitary Being or God is Real

Could it be that Newberg and d'Aquili want to allow for the possibility that God exists? Consider the following quote:

> If God exists, for example, and if He appeared to you in some incarnation, you would have no way of experiencing His presence, except as part of a neurologically generated rendition of reality. You would need auditory processing to hear His voice,

47 Newberg et al. (2001b) 35–37.

> visual processing to see His face, and cognitive processing to make sense of His message. Even if He spoke to you mystically, without words, you would need cognitive functions to comprehend His meaning, and input from the brain's emotional centers to fill you with rapture and awe. Neurology makes it clear: There's no other way for God to get into your head except through the brain's neural pathways. Even if there were a soul through which God could communicate, it would have little cognitive meaning to us without a brain.[48]

Thus, it seems that Newberg and d'Aquili want to allow for the possibility that God exists when they maintain that if God exists and wants to communicate with us, God cannot do this otherwise than by using different brain structures. Nevertheless, it is important to note that they do not maintain that science can prove the reality or existence of God. A problem is that it is not clear how Newberg and d'Aquili understand God here. As mentioned earlier, Absolute Unitary Being is a synonym for God but sometimes it is difficult to distinguish whether they regard Absolute Unitary Being or God as *sein* or as a *Wesen*. This indistinctness is obvious when they maintain that Absolute Unitary Being is real,

> so real do these experiences appear when recalled in baseline reality that they have the ability to alter the way the experiencers live their lives [...] [T]he word *real* is not here used poetically or metaphorically. It is used in the same sense as in the utterance "This rock and this table are real."[49]

Now Newberg and d'Aquili support their view by referring to plenty of evidence for its existence in cross-cultural religious literature and the testimonies of living mystics and other people who experienced Ultimate Reality, (among them the subjects of the neuroscientific studies of Newberg and d'Aquili) regardless of which religion they belong to.

These people, they say, remember the Ultimate Reality experience with the same degree of clarity and sense of reality as they remember everyday events of the past, something which is not the case with hallucinations, among other experiences. As a matter of fact, where Ultimate Reality experiences are remembered as *superior* to baseline reality, dreams and hallucinations are remembered as *inferior* to baseline reality.[50] That Absolute Unitary Being is real is so, Newberg and d'Aquili maintain, because their neuroscientific experiments have confirmed the material basis for Ultimate Reality by showing that at the time of the Ultimate Reality experience, there occurs a total deafferentation of the left and right posterior superior parietal lobes, which means that the brain is empty of all information normally processed by the brain. The ensemble of the three arguements, (1) plenty

48 Newberg et al. (2001b) 37/183: Note 1.
49 d'Aquili/Newberg (1999) 192.
50 d'Aquili/Newberg (1999) 192.

of evidence in cross-cultural religious literature, (2) testimonies of living mystics and other people who experienced Ultimate Reality and (3) the confirmation of their neuroscientific experiments, seem to be a sufficient reason for Newberg and d'Aquili to assert that Ultimate Reality (God) exists.

1. There is plenty of evidence of Ultimate Reality (God) experiences in literature
2. There are also plenty of testimonies of people living today
3. Neuroscientific experiments have confirmed the material basis for Ultimate Reality (God) experiences
4. *Therefore*, Ultimate Reality (God) is real

Whether or not the conclusions follow from the premises will be considered when I evaluate their work in the next chapter. Newberg and d'Aquili add that the "realness of Absolute Unitary Being is not conclusive proof that a higher God exists, but it makes a strong case that there is more to human existence than sheer material existence".[51]

5.3.4 Mind and Brain Relationship

How we should comprehend Ultimate Reality depends on how we approach the relation between the physical and the mental. According to Newberg and d'Aquili, "there is a strange theological conclusion to be drawn from the fact that individuals and cultures have an irreducible choice whether 'external' reality or 'subjective' consciousness is [ontologically] primary".[52] If we regard material or external reality to be ontologically primary, Ultimate Reality becomes something that is generated by (and not correlated to) the deafferented left and right PSPL. An extreme consequence of this view would be that nothing can exist over and above material aspects of life, which implies that religious experiences are totally reduced to brain activity. Newberg and d'Aquili reject such ontological reductionism because of that consequence and because they search for correlations instead. Needless to say this conclusion will also be further evaluated.

However, they continue, if we instead regard subjective awareness as ontologically primary, we must "conclude that Pure Awareness represents

51 Newberg et al. (2001b) 172. It is interesting to compare their conclusion to that of Patricia Churchland who maintains that neuroscientific investigations of religious experiences drain probability from the hypothesis that such experiences are experiences of Ultimate Reality or God.
52 d'Aquili/Newberg (1999) 193.

absolute reality, in itself neither subjective nor objective, but from which both subjects and objects are derived".[53] In *The Mystical Mind,* d'Aquili and Newberg put it "one can conclude [...] that God (Absolute Unitary Being or Pure Consciousness) generates the world (including the brain) and subjective experience itself".[54] Thus Newberg and d'Aquili theorize that the latter approach requires that both individual subjective awareness and external material reality are derived from Pure Awareness. This means that besides accepting that subjective awareness is ontologically primary (which implies every possible state of awareness, of which one is Ultimate Reality or Pure Awareness) we also have to accept that the non-physical *created* all that is physical. Newberg and d'Aquili do not develop this approach further and we do not know whether they consider it from a neurological point of view, from a philosophical or ontological one. However, what they do say is that "[a]s scientists, we have chosen to place our 'act of faith' consciously and explicitly in the priority of external reality. We [Newberg and d' Aquili] feel that this approach allows for better science, although it may not be best in other aspects of life".[55]

Thus, Newberg and d'Aquili choose the first approach to reality in order to study the physical part of life because contemporary scientific methods are based on this approach. However, they choose the second approach when they want to study other aspects of life such as awareness, love and faith for example, because then they need other methods than the ones science can place at their disposal. For instance, if Newberg and d'Aquili had not made the assumptions they did in addition to their neuroscientific research on religious experience they would not have been able to come to the conclusions they did. They cannot draw such conclusions on a neuroscientific basis alone. This implies that they have an integrated approach to the problem of subjective awareness and material reality.

They approach reality in this way on account of how they consider the relation between the brain and the mind: Brain and mind are two sides of the same coin. Newberg and d'Aquili consider the relation between the brain and the mind as follows:

> [W]hile we do agree that the mind is derived from the brain, we also believe that the interaction is much more complex and intriguing. [...] [our hypothesis] specifically holds that "mind" and "brain" are two views of the same reality – mind is how the brain experiences its own functioning, and brain provides the structure of the mind.[56]

---

53 d'Aquili/Newberg (1996) 248. Pure Awareness is another term they sometimes use for AUB.
54 d'Aquili/Newberg (1999) 193.
55 d'Aquili/Newberg (1999) 190.
56 Newberg et al. (2001b) 183.

Furthermore, they write, "mind and brain are simply two different ways of looking at the same thing".[57]

Thus to Newberg and d'Aquili, material and subjective awareness are two sides of the same reality even though material reality is ontologically primary. Furthermore, as they consider mind and brain as complementary, they also consider the material and non-material reality as complementary, which their studies give proof of. But how should we understand this complementarity?

Let us analyse the concept a little further and account for how complementarity can be understood and which understanding should be accepted for our purpose.

A first way to understand complementarity is as *completing something*. Something is not complete without the different complementary parts. The black keys of the piano are a complement to the white keys of the piano in order to be able to illustrate the half tones in a piece of music. Without the black keys (or the white keys) the piano player would not be able to perform the piece of music in accordance with the composer's intention. This is the most usual understanding of complementarity.

Secondly, complementarity can be understood as different elements emerging into something new. For instance, when I blend red paint with white paint I will have a new colour, pink. In this case I cannot return to the original elements.

Thirdly and similar to the first definition, but not quite the same, is to understand complementarity as serving to fill out something, such as, adding a subordinate clause to make a sentence more attractive or clear. The difference with the first option is that the piece of music needs both the white and black keys but the sentence does not really need the subordinate clause.

A fourth way to comprehend complementarity is as two items supplying each other's lack. Friends, for instance, can be such complements to one another.

Finally, it can be understood as two sides of the same coin. It is understood in this way in quantum mechanics, more precisely, in the wave-particle complementary principle which states that light shows itself at one time as (collapsing into) waves and at another time as particles. It is in the latter way that Newberg and d'Aquili understand complementarity. They write:

> In this model, light can sometimes appear to function as a particle and sometimes can appear to act as a wave, depending on how one tries to measure and analyze the light. [...] Similarly, depending on what one is looking for, one might describe thoughts and

57 d'Aquili/Newberg (1999) 21.

feelings as originating from the structure of the brain or from the product of that structure – the mind.[58]

Brain and mind are thus assumed to be two sides of the same coin, because "the brain is a collection of physical structures that gather and process sensory, cognitive, and emotional data; the mind is the phenomenon of thoughts, memories, and emotions that arise from the perceptual processes of the brain".[59]

Thus, according to Newberg and d'Aquili, the brain is associated with receiving input from the external world and with processing this input, while it is through the mind, that correlates with structures of the brain (in a way they do not clarify since science probably has no means of doing so), that this input makes sense to us:

[W]hen we look at our mother, our brain receives the image through our eyes; [the mind] analyzes the colors and shapes and interprets the image of "mother"; informs us that the image is, in fact, our mother; decides on a specified type of emotion or behavior towards this image; and initiates, generates, and determines that behavior.[60]

Thus, whether we study the mind or the brain depends on what we want to know. In other words, if we were to analyse what happens in the brain when we see our mother, we would consider the neurological structures underlying the reception of the image of the mother: i.e., the retina, the visual brain areas etc. Thus we would perform a neurological analysis. If, on the other hand, we were to analyse what happens in the mind when seeing our mother we would consider our subjective feelings and thoughts associated with the seeing of our mother. Thus we would perform a psychological analysis. The first analysis will result in empirical data while the second will give us phenomenological information. The first analysis could be performed by way of neuroimaging, while the second analysis could be performed by way of, for example, psychoanalysis.

According to Newberg and d'Aquili, and if I have understood them correctly, these two analyses are complementary because the brain and the mind are considered to be complementary to one and the same person.

We shall now take a closer look at the consequence of regarding the material as ontologically primary. According to Newberg and d'Aquili, a consequence is that God or Ultimate Reality has to be *hard-wired* into the

58 d'Aquili/Newberg (1999) 21–22.

59 Newberg et al. (2001) 33.

60 d'Aquili/Newberg (1999) 22–23. Here we have an example of a philosophical problem caused by using a language that personifies the brain, namely Our brain informs us: who is us and who is the brain and how can the brain inform us? I shall not analyse this further nor will I comment on every such problem but I want to make a remark that such kind of philosophical problems may occur.

brain.[61] This is because, as mentioned earlier, in this case all experiences of God or Ultimate Reality are generated by the brain. Now, in its natural context, i.e., computer-sciences, hard-wired means that a particular feature or a possibility is constructed directly into the hardware, on the printed circuit board. In other words, the computer is designed with this specific feature or possibility as integral to it. This feature or possibility is not added afterwards nor is it a program component (i.e., software component). This has consequences for how Absolute Unitary Being should be understood.

Firstly, if Newberg and d'Aquili understand God as an Ultimate Wesen then *God has to be hard-wired* means that God, or the possibility for experiencing God, is constructed into our brain from the birth of the human race. Understanding God as such implies that God *created* humankind in such a way (for purpose of survival or other).[62]

Secondly, if Newberg and d'Aquili understand God or Ultimate Reality as Ultimate Sein, the view that God is the absolute ground of everything that is and ever will be, *God is in the brain* in exactly the same way as in everything there is and ever will be. Accepting this implies that, if the brain should generate God experiences, it follows that the brain generates all experiences and nothing could exist independently of the brain.

On the other hand Newberg and d'Aquili write that if subjective awareness is considered to be ontologically primary, we have to admit that certain individuals experience God as a primary epistemic and ontological state.[63] The reason why is similar to the reason given above, namely, once subjective awareness is accepted as ontologically primary, one has to accept all the attributes of subjective awareness, including all experiences on the Aesthetic-Religious continuum. Even though God can be experienced only by way of the brain, God is not caused by the brain. The brain is considered to be the instrument by which God becomes accessible for us, regardless of whether we understand God as Ultimate Reality, *Ultimate sein* or *Ultimate Wesen*.

Newberg and d'Aquili also claim that experiences of Ultimate Reality are rather rare but they emphasise other religious experiences described by many persons. Examples of such religious experiences which are frequently described by people are (1) having met an angel or having been saved by an angel, (2) having experienced a revelation by, for example, the Virgin Mary, who brought a special message for the receiver of the revelation or his or her environment, (3) perception of signs which the receiver interpreted as signs from God and which had a particular purpose in the life of the receiver of the sign, etc. These religious experiences range within the

61 d' Aquili/Newberg (1999) 180–186.

62 Remember Persinger's hypothesis.

63 d' Aquili/Newberg (1999) 187–190.

realm of religious experiences, somewhere on the Aesthetic-Religious Continuum. How they rank religious experiences in relation to each other on this continuum is not clear.

What we should keep in mind, however, is that Newberg and d'Aquili consider it wrong to dismiss descriptions of religious experiences in the world's religious and mystical literature as the "silly imaginings of religious nuts".[64] On the contrary, they write that

> [t]he road to God is paved with many stones: metaphor, poetry, music, ritual experiences, prayer, and meditative experiences. [These are] indeed culturally conditioned. But relative as they are, they point to that which is "really real". Metaphor does not point to more metaphor, nor does religious poetic language point to more poetry. That [Absolute Unitary Being or God] which gives religious poetry and metaphor its meaning is real in an absolute, ultimate, and unconditioned way, as the experience AUB [or God] clearly demonstrates.[65]

Thus Newberg and d'Aquili consider that there are several ways to explain religious experience but that all explanations, however different they may be, are alike in the sense that they are about an experience of something which the experiencer remembers and describes as more real than everyday reality. Newberg and d'Aquili compare the descriptions of the experiencers of what they experience with poetry and maintain that poetry does not point to more poetry but to something the poem is all about. Consider the following poem by William Butler Yeats:

The Sorrow of Love

The quarrel of the sparrow in the eaves,
The full round moon and the star-laden sky,
And the loud song of the ever-singing leaves,
Had hid away earth's old and weary cry.
And then you came with those red mournful lips,
And with you came the whole of the world's tears,
And all the sorrows of her labouring ships,
And all the burden of her myriad years.
And now the sparrows warring in the eaves,
The curd-pale moon, the white stars in the sky,
And the loud chanting of the unquiet leaves,
Are shaken with earth's old and weary cry.

---

64 d'Aquili/Newberg (1999) 206.
65 d'Aquili et al. (1999) 210.

Following Newberg and d'Aquili, then, the poem does not point to another poem but to the cause that makes Yeats experience the pain of being in love. As it was very difficult or even impossible to account for his feelings in the everyday English language, he takes to the language of art. So, according to them, it is descriptions of religious experiences which point to the cause of the religious feelings of the experiencer. Some might argue that the poem could also refer to a fictive picture of love created by poetry itself and that such could also be the case for the referent of a religious experience, i.e., created by the specific religion itself. However, in that case we are not talking about trying to put an indescribable experience into words. Might this perhaps be a reason why Newberg and d'Aquili put romantic love experiences between aesthetic and religious experiences?

5.4 Summary and Conclusions

Newberg and d'Aquili maintain that one has to consider both the material and the spiritual side of the world and they believe that neuroscientists may significantly contribute to the process of understanding religious experience. Two important goals for their research were especially pointed out, (1) they wanted to clarify which element(s) if any, is (are) invariant across the different religions and (2) they wanted to find specific correlations between religious experience and neural activity in brain functions and structures. The experimental study may have had a third goal, namely, to allow them to confirm, confute or adjust the results from their previous research on religious experience obtained through meditation.

Their neurological models concern different acts of meditation. We analysed their studies of passive meditation or via negativa and active meditation or via positiva. To facilitate the construction of their neurological models, Newberg and d'Aquili also designed a scale, called the Aesthetic-Religious Continuum, on which they ranked different types of experience ranging from basic experiences (being hungry), through the realm of aesthetic experiences (watching Magritte's paintings) into the realm of religious ones (perceiving God in nature) and ending with the ultimate religious Absolute Unitary Being experience. They rank the experience of romantic love between the Aesthetic and the religious experience to which they attribute a transcendent notion.

There are two different ways in which we could understand Absolute Unitary Being. A first possibility is to regard Absolute Unitary Being as Ultimate Reality as I intended, namely, as the absolute ground of everything that is (and ever will be). Comprehended as such, being means to be, sein in German. Thus, Absolute Unitary Being then means Absolute Unitary Sein.

The second possibility is to understand Absolute Unitary Being in terms of a Being, ein Wesen in German. Hence, Absolute Unitary Being means an Absolute Unitary Wesen. Since they do not intend to affiliate Absolute Unitary Being with specific religions (one of their aims is namely to find an invariant element across all religions), I suggested that they understand Absolute Unitary Being as Ultimate Sein and not as an Ultimate Wesen. Unfortunately they do not always use the concept in a consistent manner.

Newberg and d'Aquili regard Absolute Unitary Being as the ultimate experience that can be achieved. The meditators who experienced Absolute Unitary Being describe the experience as follows: it does not include any perception or awareness of discrete beings, nor of space or time, there is no self-other or self-world distinction. Newberg and d'Aquili hold that there cannot be two different Absolute Unitary Being experiences and base this assumption on their neurological models for the process of meditation. This experience should correlate with a total deafferentation of both the right and left posterior superior parietal lobes (PSPL). Unfortunately, they could not confirm this hypothesis by way of their neuroimaging study. The experiment only resulted in an overall tendency in that direction. In their research report, Newberg and d'Aquili did not relate specific neural activity during the meditation/praying act to the individual persons. Therefore we do not know whether the results are related to Tibetan meditation or Franciscan praying or to both.

Newberg and d'Aquili assert that the Absolute Unitary Being experience, at the very moment of the experience and before the experiencer puts the experience into words, is similar for all. They reason that because Absolute Unitary Being correlates with the total deafferentation of both the left and right PSPL and because there cannot be two different totally deafferented PSPL (brain-imaging of various deafferented brain-structures show exactly the same neural activity), there hence cannot be two different Absolute Unitary Being experiences.

Newberg and d'Aquili also consider all experiences to be equal, neurologically, and maintain that if religious experiences are nothing but neurochemical flux, then all experiences are nothing but neurochemical flux. They compare the experience of eating an apple pie with Absolute Unitary Being experience or God. However, it seems that Newberg and d'Aquili want to allow for the possibility that Absolute Unitary Being or God exists. An Absolute Unitary Being experience is real and what they intend by real, is real in the way that this computer and this lamp are apprehended as real. However, even if Absolute Unitary Being is considered as a synonym for God, Newberg and d'Aquili do not always apply Ultimate Reality or God in a consistent manner. Sometimes it is difficult to distinguish whether they regard Absolute Unitary Being or God as Absolute Unitary Sein or as a an Absolute Unitary Wesen.

Newberg and d'Aquili maintain that if God exists and wants to communicate with human beings, God will have to use the brain's neural pathways. Furthermore, according to them there is plenty of evidence of experiences of God in literature. There are also plenty of testimonies of people living today who have experienced God and neuroscientific experiments have confirmed the material basis of the God-experiences. Therefore, God or Ultimate Reality is real.

How we understand Ultimate Reality or God depends on how we approach the relation between the physical and the mental. If we regard the material, also called external reality, as ontologically primary, Ultimate Reality becomes something that is generated by the deafferented left and right PSPL. An extreme consequence of this view would be that nothing can exist over and above the material aspects of life, which implies that religious experiences are totally reduced to brain activity. Newberg and d'Aquili reject such ontological reductionism.

If, instead, we consider subjective awareness as ontologically primary, we shall have to accept that both individual subjective awareness and external material reality originate in Ultimate Reality. The non-physical created all that is physical.

Newberg and d'Aquili chose the former approach because of today's scientific methods and chose the second approach when they wanted to study non-physical aspects of life. They approach reality in this way because they consider the mind and the brain as complementary, in a quantum mechanical wave-particle sense.

Whether we study the brain or the mind depends on what we want to know. If we want to analyse what happens in the brain when we look at something we must investigate, for example, the visual structures of the brain. If, on the other hand, we want to analyse what happens in the mind when we look at that something, we must investigate the subject's thoughts and feelings associated with seeing that something. The former will be a neurological analysis, the latter a psychological one. The analyses are considered to be complementary, because the mind and the brain are considered to be complementary in one and the same person.

Regarding material or external reality to be ontologically primary means that we encounter the consequence that Ultimate Reality or God has to be hard-wired into the brain. This consequence can be understood that God is nothing but a ghost in the brain. However, this depends on how we understand God. If we understand God as Ultimate Reality being the absolute ground of everything that is and ever will be (Ultimate Sein), then God is already in the brain. If, on the other hand, we understand God as ein Wesen, then *God has to be hard-wired* means that God or the possibility of experiencing God is constructed into the brain from the birth of the human race. It means that God created humankind in such a way (for the purpose of survival or other).

If subjective awareness is considered to be ontologically primary, God is not caused by the brain: rather the brain is the instrument by which God becomes accessible for us, regardless of whether we understand God or Ultimate Reality in the sense of sein or Wesen.

Newberg and d'Aquili's research on religious experience is not without problems and needs evaluation. This will be done in the next chapter.

6 Evaluating Newberg and d'Aquili's Research

Having analysed Newberg and d'Aquili's research on religious experience obtained by way of different meditation techniques, we can now begin the evaluation of their work. I shall investigate how the empirical study is performed from a scientific point of view and explore whether, and if so to what degree, their experimental study measures what it purports to measure.[1] Thereafter, the relation between the results of the experimental study and the neurological models for religious experiences obtained by way of meditation will be considered. The following questions were asked: Were Newberg and d'Aquili able to find the invariant element across the religions which they were looking for? Is it the case that all experiences are equal neurologically and should we accept Newberg and d'Aquili's suggestion that God or Absolute Unitary Being is real or really exists? I shall also look more closely at the implication of one of the results of the neuroimaging study, namely, that meditators and non-meditators had different thalamic values at baseline.

6.1 The Scientific Validity of the Study

Newberg and d'Aquili maintain that neuroscience has significant potential in helping to understand at least some of the many varieties of religious experiences. However, they assert that it is also important to consider the limitations of neuroscientists, especially concerning their epistemological and ontological possibilities for contribution to this understanding. Nevertheless Newberg writes, "our goal was not only to develop as detailed a theory as possible but also to explore all of the implications for theology, philosophy, and the sciences".[2] Newberg and d'Aquili maintain that one has to be careful in drawing conclusions from neuroscientific results.

As mentioned in the previous chapter, at least three reasons may have guided Newberg and d'Aquili in their neuroscientific research on religious experience: (1) To clarify which element(s) if any, that is (are) invariant across the different religions. (2) To find the specific correlation between religious experience and neural activity in brain functions and structures.

1 Newberg et al. (2001a).
2 Newberg (2001) 501.

(3) To allow them to confirm, confute or adjust the results from the neurological models for religious experience.

As seen earlier in the analysis of their work, Andrew Newberg and Eugene d'Aquili chose to study religious experiences obtained by way of different meditation techniques, because spontaneous religious experiences are more complicated to use as the basis for neuroscientific study. We saw earlier that one of the problems is that it is not known which factors trigger spontaneous religious experiences, while persistent training and endurance are known to be the factors that bring about religious experiences obtained, among other methods, through meditation. Furthermore, the latter religious experiences can be repeated and tested in the way science prescribes because the act of meditation can be performed in laboratory settings. Also mentioned earlier, spontaneous religious experiences are more complicated to use for designing models for the correlation between the brain and the experiences than those obtained through meditation because of the repeatability of the latter and its accompanying documentation. Thus, they satisfied the criteria of clinical neuroscience.

Newberg and d'Aquili chose their participants, i.e., their sample, from among Buddhists and Christian nuns who had meditated on an intensive and regular basis for at least 15 years. In this way they were able to study the experiences of advanced meditators from both theistic and non-theistic religions, i.e., the population. However, it is questionable whether the sample is too small considering its purpose (to find the correlation between the brain and religious experiences obtained by way of different meditation techniques used in theistic and non-theistic religious traditions) since it did not include meditating participants from, among others, the Muslim or Hindu traditions, who also belong to this population. Thus, one might argue, Newberg and d'Aquili did not fulfil the requirements of what an adequate sample should be. Consequently the sample may not be representative for the population of study.

However, Newberg and d'Aquili could defend their choice of sample by saying that adding members of other religions would not make any significant difference to the representativeness of their sample because their population of study reflects advanced meditators belonging to the theistic and non-theistic religions, and their sample reflects a proper subset of this population. They could argue that their study does not concern advanced meditators belonging to different religions *per se* and hence the criticism concerns a different kind of study.

Nevertheless, if Newberg and d'Aquili's sample is to be representative for the population of study they need to consider both men and women of different ages. This is a deficit in their research according to James Ashbrook who refers to previous EEG studies showing different topographic maps of brain electrical activity in females and males.[3] The study performed by Charlotte Smith showed that men had the highest amplitude in the frontal cortex and women in the central cortex. The EEG study also showed that women have more bilateral use of the hemispheres than men.[4] Ashbrook concludes that men should be "oriented more to a cognized, dichotomized, and impersonal environment" and women "to an operational, connected, and interpersonal environment".[5] He argues that as long as Newberg and d'Aquili do not consider the differences between men and women, their "scheme lacks universal relevance".[6] Even if Newberg and d'Aquili's sample meets the gender requirements, i.e., the sample includes both men and women, they do not refer to it in their discussion of the results. What would also have been desirable would have been to choose a group of persons consisting of a broader age-range. All meditators were between 38 and 52, which is not an optimal representative sample of the population of meditating people. Surely, there are advanced meditators who are younger than 38 and who are older than 52.

Another remark is that the sample contained too few items, i.e., persons who practise meditation. This might endanger the validity of Newberg and d'Aquili's experiment, as well as the reliability of the results. However, Newberg and d'Aquili comment on this in connection with their discussion of one of their results: "This is an interesting finding that needs to be explored by examining a larger sample of subjects".[7] Hence, the results of their study would probably have been more accurate if they had used a larger sample.

Concerning how the experiment was performed, I have two remarks. I do not believe that my considerations would change anything concerning the scientific validity of the experiment but they are nevertheless interesting, because they might improve the results.

The first one is that I would have liked Newberg and d'Aquili to have given a cognitive task to the non-meditating control group. For example, I would have let them concentrate on a poem (repeatedly reading or rehearsing it). This is similar to how Nina Azari performed her neuroscientific experimental study of religious experience. The aim of Azari's studies is different from Newberg and d'Aquili's and her sample does not include meditating people. She wanted to show that religious experiences correlate with the cognitive structures of the brain and not with the limbic system as

3 Ashbrook (1993) 242.
4 Ashbrook (1993) 242.
5 Ashbrook (1993) 242.
6 Ashbrook (1993) 242. Gender issues are, admittedly interesting but not the focus of the present research project. Hence, I will not consider it further at this time.
7 Newberg et al. (2001a) 121.

other neuroscientists had suggested. Therefore, the participants of her study had to do cognitive tasks. More precisely, the religious people had to read a specific psalm from the Bible and later memorise it, while a non-religious group had to read a famous poem and later memorise it. The control group had to read a certain passage from the telephone book and later memorise it. All were scanned after a certain time by PET imaging. Azari concluded that "religious experience may be a cognitive process, which nonetheless, feels immediate".[8]

To return to Newberg and d'Aquili, in my opinion, if they had given the non-religious participants a cognitive task, they would have been able to compare the scans of the brains of the meditators and the non-meditators also at a later moment. They could have compared those taken at the moment when the meditator pulled the twine with those taken at the moment when the non-meditator ended the cognitive task according to the instruction that he or she had received. Now they were only able to compare the scans of the brains of both groups at baseline. Since meditation and prayer also correlate with these structures of the brain that have to do with cognition, they might have discovered interesting differences and/or similarities between the group performing a specific cognitive task and the group performing meditation.[9]

Secondly, I would also have preferred Newberg and d'Aquili to be more careful in distinguishing the types of meditation when they took the images of the brain of the meditators. I argue that it would have been very interesting to learn whether and to what extent the neurological models for religious experiences obtained by using different meditation techniques correspond with the neuroimaging data. Do the results of the pictures of the brain taken by the SPECT on a person performing the via negativa meditation, showing the neural activity that correlates with her or his religious experiences, correspond to the neurological model of which parts of the brain should correlate with the religious experiences during such a meditation process? For instance, can the IPL only be associated with the via negativa and not with the via positiva and, similarly, can the ITL only be associated with the via positiva and not with the via negativa?

Concerning the method and equipment Newberg and d'Aquili used, I suggest that they chose the SPECT in preference to the fMRI and PET for good reasons. Firstly, Newberg and d'Aquili considered other neuroscientific studies of religious experience, such as EEG studies of the brain of meditating people, when they chose their own research instrument. Secondly, they chose to use the SPECT as an imaging technique because even though the

8 Azari (2001) 1649.

9 Observe that Newberg and d'Aquili call meditation "the Complex Cognitive Task" Newberg et al. (2001a). However, Newberg and d'Aquili's research results also show that religious experiences also involve limbic activity.

fMRI has improved resolution compared to SPECT, the meditators found it impossible to meditate because of the noise of the fMRI. Thirdly, SPECT has the advantage that the radioactive tracer that is injected before and during the act of meditation and which is carried by blood flow, almost immediately locks into brain cells and remains there for hours (up to 24 hours). This allows the researchers to scan the person for up to 24 hours after the injection of the radioactive tracer, which means that the investigators did not need to *disturb* the person during the act of meditation in order to make a diagnosis. Another advantage of SPECT is its fast uptake values allowing the investigators to observe the brain at the very moment of the experience.

6.2 Comparing Experiment and Models

Newberg and d'Aquili hoped that their models of how different brain functions and structures should correlate with different religious experiences would be experimentally confirmed, if not completely, at least to a significant extent. Was this the case?[10]

One would expect that because Newberg and d'Aquili studied persons who meditate and pray (probably using both the via positiva and the via negativa meditation technique) by way of neuroimaging technology, we should be able to compare the results of their experimental research with the neurological models. However, this is rather difficult to do. One reason why it is difficult to compare the results of the experimental research with the neurological models is that Newberg and d'Aquili, in their neuroimaging study of persons who meditate and pray, did not clearly distinguish between the types of meditation used when they scanned the brains of the meditators. In other words, their experimental study does not explicitly tell us whether the meditators practised the via positiva or the via negativa meditation technique.

Other reasons are: Firstly, the models do not have any hypotheses about increased or decreased neural activity in specific brain structures, because such can only be shown by way of neuroimaging technology. The neurological models merely suggest how the process of a meditation section should be neurologically, i.e., which brain structure(s) should stimulate which in the process. Secondly, since the subjects of the experimental study are only SPECT-scanned twice, the first time at base-line and the second time when they feel that they reach their most intense moment of meditation, there is no scheme that follows the meditation process as is presented in the theoretical models.

10 d'Aquili (1993b), d'Aquili/Newberg (1999), Newberg et al. (2001a).

However, some results of the experimental study confirm the hypotheses of the neurological models. (1) The frontal lobes, which have to do with cognition, in particular the prefrontal cortex (PC), had significant increase of neural activity. Altered neural activity was suggested in the neurological models. (2) Decreased neural activity was found in the PSPL during meditation. However, as mentioned earlier, the results of the neuroimaging study did not show a total deafferentation as was suggested in the models. (3) The occipital area, which is involved in visual perception, had increased neural activity during meditation. It was suggested that the via positiva meditation technique would result in altered activity in the visual parts of the brain. (5) The thalamus, part of the limbic system which is the integration centre for sensory information, had decreased neural activity during meditation. However, the experimental study also showed that the meditators had a different thalamic laterality index at baseline than the non-meditating control group, which was not hypothesized in the neurological models. On the other hand, these models associate more brain functions and structures with the different meditation techniques than was shown in the result of the experiment. Whether these agreements are significant, as Newberg and d'Aquili wished, is difficult to say.

Once more, it is unfortunate that in their research report Newberg and d'Aquili did not relate specific neural activity during the meditation/praying act to the individual persons. Therefore we do not know whether the results are related to Tibetan meditation or Franciscan praying or to both.

The results of the experimental study tell us whether the hypotheses were confirmed or not and they show which brain structures and functions had extraordinary neural activity at the most intense moment of meditation and prayer. However, the results do not reveal anything about the religious experiences per se, about the quality of the experiences, about the core characteristics of the religions to which the meditators belonged, nor about the reality status of the experiences, i.e., whether the experiences are what they are said to be or are illusory. Their neurological models may be more illustrative concerning what types of experience may occur during different meditation processes, but also these do not, by themselves, tell us anything more than the correlating neural activity, which already needs the cooperation of the meditator, or knowledge of such a process.

6.3 Invariant Element across the Religions?

How should we apprehend Newberg and d'Aquili's conclusion that the Absolute Unitary Being experience could be the invariant element across all religions? The reason why they believe this is so, is because according to them, all meditators, regardless of meditation technique and regardless of which religion they belong to, experience Absolute Unitary Being as the ultimate experience of meditation.[11]

In my view, Newberg and d'Aquili's idea has its origin in their neurological models. If we consider the model of the via positiva meditation process, we see that the meditators, at a given moment, experience a union with the object of meditation or prayer which lasts as long as they are able to focus on that object. However, we see that once they lose that concentration, a deafferentation of both the left and right posterior superior parietal lobe ought to occur and they should experience Absolute Unitary Being. Newberg and d'Aquili then refer to the neurological correlation of the Absolute Unitary Being experience, namely, the total deafferentation of both the left and right PSPL and, because neurologically, there cannot be two different deafferented posterior superior parietal lobes, they maintain that there cannot be two different Absolute Unitary Being experiences. As mentioned in the analysis above, I believe that they reasoned as follows:

1. If there is an Absolute Unitary Being experience, then there is also a total deafferentation of the both the PSPL.
2. Two different total deafferentations of both the PSPL are not possible
3. *Therefore*, there cannot be two different Absolute Unitary Being experiences

However, as mentioned earlier, this is problematic from a neuroscientific point of view, because some neuroscientists maintain that a total deafferentation of the left and right PSPL is simply not possible. For instance, Mogens Dalby maintains that the condition of persons who suffer from two-sided infraction in the PSPL (which is a partial deafferentation of the left and right PSPL), is such that they lack bodily perception, they cannot follow a line of arguement, they do not recognise the objects which are placed in their hands or they cannot walk, etc. Hence, they do not seem to have the same control over what happens as the meditators seem to have. Dalby argues that one cannot compare the neurological condition of these persons with the neurological condition of meditators experiencing Ultimate Reality.[12] The neurological condition of the persons suffering from damage to the PSPL is regarded as a *proper* deafferentation of that brain area, (the condition is invariable), while the neurological condition of the meditators is considered to be a modulation of the functions of that brain area. We saw earlier that proper deafferentation means that all neural input or flow into a

11 Newberg et al. (2001b) chap. 6/9.
12 From my correspondence with Dr. Dalby.

certain brain structure is *cut off*.[13] This remark is related to the decrease in neural activity (the blood flow decrease) that Newberg and d'Aquili see in the PSPL area. It does not have to imply that the PSPL is completely cut off from its normal input. These neuroscientists make a distinction between the "*shutting down* or *cutting off* of the [PSPL] area. [...] Shutting down neural activity is known as *inhibition*, and inhibition of a brain area may result in decreased blood flow to that area".[14] Thus, Newberg and d'Aquili should have been more precise in what they meant by total deafferentation of the left and right PSPL, because their models seem to "require that the [PSPL] area remains active while it is cut off from its normal neural input",[15] but they probably mean *while it is shut down from* its normal neural input. Newberg and d'Aquili could have added that during the experience of Absolute Unitary Being, the left and right PSPL are deafferented but not *deefferented*, i.e., even if there may not be information to generate a sense of orientation, information on how to interpret this lack of orientation will still be sent to adequate parts of the brain. While deafferentation refers to the inhibition or blockage of input into a brain structure, deefferentation refers to the inhibition or blockage of output from a brain structure.

Those of us who are familiar with religious studies or theology know that their idea of an invariant element across the religions is not new. Newberg and d'Aquili are not alone in the search for a common core for all religions. For instance, Stefan Einhorn presented the idea of the existence of a concealed God in all religions as the core of unity of religions.[16] Einhorn explains his idea by way of analogies. Let me reconstruct one of his thought experiments.

Eating an apple requires our five senses. However, were we blind we would only have four senses with which to experience eating the apple. Assume, then, that all people were blind. This would imply that there would probably not be a language for, for example, colours. Assume further that Mary has an intuitive notion that colours exist and that she wants to scientifically prove this. How would she proceed? She would not find any adequate method to help her explain the term colour, because the language does not include that word. Thus, Mary would probably explain colours within the framework of her cultural-religious-personal reference, let us call

13 From my correspondence with Dr. Dalby.

14 Spezio (2001) 482.

15 Spezio (2001) 482.

16 Stefan Einhorn is professor of molecular oncology and pathology and chairman of the department of oncology-pathology at the Karolinska Institute, Sweden. Einhorn, like Newberg, is a medical scientist, working in a clinical environment and interested in the spiritual side of human beings. Where Newberg wants to understand the correlation between the brain and religious experiences, Einhorn wants to understand how everything is connected. His approach is a scientific approach, searching for logical conclusions that bind everything together.

it *A-reference*. Assume that another person in another part of the world, let us call him Peter, has his reasons to prove that colours exist. Obviously, he will encounter the same problems as Mary. However, Peter may not explain colours in the same way as Mary, but may explain them according to his own frame of reference, say *B-reference*. The point Einhorn wants to make is that both Mary and Peter describe the same *concealed colours.*[17] Similarly, he argues, religious people describe the same *concealed God.* The concealed God as Einhorn sees it may be understood both as *Ultimate Reality Sein* and *an Ultimate Wesen.*

However, we should be careful not to draw the conclusion that all religious experiences are reducible to cultural-religious-personal issues. Gregory Peterson, for instance, draws our attention to the problems found in such an approach. He maintains for instance that "[t]he colors I experience are not limited to my relatively impoverished color vocabulary, and no amount of reading about parental love can equate with and prepare one for the actual blend of emotions".[18]

Also philosophers of religion have searched to find an invariant element across the religions. One of them is John Hick. In his pluralist hypothesis, Hick designated *the Real* as the invariant element across the religions. He distinguishes between "The Real *an sich* and the Real as variously experienced-and-thought by different human communities."[19] According to Hick, the Real in itself cannot be humanly experienced. The Real in itself is manifested in (experienced through) different faces or *personae* such as God, Ultimate Reality, Brahman, Allah, etc. which are the objects of worship or contemplation within the different religious traditions. The Real is "a necessary postulate of the pluralistic religious life".[20] This postulation is necessary because without it, it becomes difficult (if not impossible) to equate the authenticity of Ultimacy experiences of different religious traditions. The Real in itself is neither personal nor impersonal but *becomes* personal or impersonal in relation to the experiencer's cultural and religious affiliation as they recall their religious experiences. Thus, these experiences are tied to the cultural-religious-personal background of the experiencer, which should allow us to interpret the Real as Ultimate Sein but also as an Ultimate Wesen. Could these pluralistic ideas be compared with Newberg and d'Aquili's? According to Gellman, Hick's pluralist hypothesis could

> square d'Aquili and Newberg's theory with the pluralist hypothesis, by maintaining that cultural learning determines the way religious experience plays itself out in the

17 Einhorn (2002) 1998.
18 Peterson (2002) 242.
19 Hick (1989) 235.
20 Hick (1989) 249.

brain. Theistic acculturation will cause the brain to reinforce lateral hypothalamic discharge, producing a theistic experience, while non-theistic religious training will bring about dominance of the ventromedial hypothalamic structures, and a corresponding non-theistic experience.[21]

Without going into the neurological part of this quote, what Gellman means, I believe, is that Hick's pluralist hypothesis would settle Newberg and d'Aquili's theory in that the cultural-religious-personal background of the experiencer may not only be significant for the interpretation of the religious experience by the experiencer but even for the religious experience itself and its neurological correlations.[22] Once more, Peterson agrees that there is a correlation between the cultural-religious-personal issues and (religious) experiences but rejects that they should be determined by them.[23]

The problem is, however, that Newberg and d'Aquili do not refer to the different neural pathways that correlate with the religious experiences obtained by way of different meditation techniques in their suggestion that there should be an invariant element across all religions. Their approach is different. They approach the issue from the neurological position, arguing that there is only one possible total deafferentation of the left and right PSPL, which correlates with an Absolute Unitary Being experience. All meditators, regardless of the meditation techniques used and regardless of which religion they belong to, sooner or later experience this. That is the reason why they conclude that there cannot be two different Absolute Unitary Being experiences. They maintain that Absolute Unitary Being experiences may be interpreted as "the *unio mystica*, or the experience of God. [...] as the void or Nirvana of Buddhism, or as the Absolute of various philosophical disciplines".[24]

To maintain this is problematic, because some Christian mystics, for example, would not describe their Ultimate Reality experience (Absolute Unitary Being) as an experience of nothingness, but of love. Some may say that they felt united with God, others may say that they felt as a drop of water in the ocean. Again others interpret their experience as an experience of total emptiness or nirvana, etc. Let me propose the following stories by way of illustration.

---

21 Gellman (2001) 98.

22 Peterson (2002) 242.

23 We have briefly seen the importance of the cultural-religious-personal background of the believer when we discussed Persinger's view on religious experience and the importance of hemisphericity. For instance Laughlin et al. (1993).

24 d'Aquili/Newberg (1999) 110. The quote is taken from *The Mystical Mind*. In *Why God Won't Go Away*, however, unio mystica is considered the ultimate religious experience of the via positiva meditation technique and has *lower* status than Absolute Unitary Being. See Newberg and d'Aquili, 2001, 164. Since Newberg and d'Aquili do not use the term unio mystica, I shall not refer to it again.

Imagine Sister Helen, a Christian nun, sitting on a pillow on her knees before the altar of a little chapel. Assume that she has been praying in the same position for hours, her eyes fixed on the candle light on the altar. Assume further, that she experiences feeling as one with the light of the candle she was focusing on, and after a while experiences God or Ultimate Reality. Thus, later on, she will remember this experience as having been one with the light. Furthermore, belonging to a Christian theistic religion, Sister Helen will probably say something like "the *Light* is one nature consisting of the oneness of life which is the Father, the *Light itself* which is the Son and the *glow of the Light* which is the Holy Spirit".[25]

Imagine also Rajiv, who belongs to the Hindu tradition and who also experienced God or Ultimate Reality. Belonging to the Hindu tradition, Rajiv will remember his experience as a feeling of being a passenger in his body. He will probably say that he did not feel any anxiety, doubt, fear, ambition, goal, desire, pain, weakness, struggle, unhappiness, happiness, frustration, anger, disappointment, etc. He might even say that he does not long for anything other than Brahman.

In describing the light, the Catholic nun describes her mystical experience in a way that is doctrinally and soteriologically central to the theistic (Christian) tradition. Similarly, Rajiv describes his mystical experience in a way that is doctrinally and soteriologically central to Hindu religious tradition. Surely, both the nun and the Hindu believer take the perceptual character of their experience seriously, which Newberg and d'Aquili's explanations of a total deafferentation of the left and right PSPL does not capture.

However, Newberg and d'Aquili could defend their idea by maintaining that the Absolute Unitary Being experience at the very moment of the experience and before the experiencer puts the experience into words *has to be* identical for all. They argue that the total deafferentation of the left and right PSPL which according to neuroscience, *is* identical for all such deafferentations correlates with the Absolute Unitary Being experience. For instance, according to neuroscientists, a person who suffers from such a deafferentation will lose the capacity to create three-dimensional objects, to distinguish between graspable and non-graspable stimuli and the capacity to locate and coordinate her- or himself in space. These experiences, according to Newberg and d'Aquili, are also described by the meditators.

Newberg and d'Aquili could defend their idea by comparing the brain state of total deafferentation to the physical state of vacuum. There are no two different vacuum states possible. There is nothing there. Can there be two different *nothings*? However, I would argue that brain states and physical states cannot be compared. The reason there cannot be two different states of

25 My translation from Nijs/Meyers (2001) 23.

vacuum is because physical states are ruled by *physical laws* which are universal. When A so B. Brain states are ruled by *neurological laws* but since every brain is unique the result will not only be when A then B, but also when A then B', and when A then B", etc., depending on the individual's brain.[26]

What Newberg and d'Aquili remind us of is that we should be careful to distinguish between the religious experience per se and the religious experience furnished with an interpretation. As we saw in the analysis of Persinger's work, the English language does not use different words to distinguish one religious experience from another. I have chosen the German words *Erlebnis* to represent the religious experience per se (non-interpreted) and *Erfahrung* to stand for the interpreted religious experience, a distinction that proved to be fruitful.[27] Following this description, an Absolute Unitary Being experience, as Newberg and d'Aquili intend it to be, should be understood as a religious Erlebnis (RErl), i.e., before any interpretation is made. The religious experience presented by the Christian nun Helen and the Hindu believer Rajiv is a religious Erfahrung (RErf), i.e., the religious Erlebnis (RErl) has been embellished with an interpretation.

Nevertheless, we could raise the question whether it is possible to have non-interpreted experiences at all. I wonder whether newborn babies have uninterpreted experiences. As newcomers in this big scary world they have to learn to interpret their experiences, which does not mean that newborn persons are *tabula rasa*. The idea is rather that their experiences are related to the world they come from, their mother's womb. According to recent psychological research, newborn babies actually recognise the music that their mother was listening to during the last month of pregnancy. This music is later interpreted in the baby's own way as pleasant (the baby smiles) or unpleasant (the baby cries). Foetuses interpret their mother's heartbeat, her stress, etc. which the newborn will later recognise, and again interpret as pleasant or unpleasant. Similarly, they will interpret their first experience of being hungry. A newborn baby who experiences hunger screams. This means that the little girl or boy manifests the experience directly. The question is not whether or not newborn babies interpret, rather it is how they interpret their experiences. My suggestion is that the foetuses and the newborn baby's interpretations are not primary cognitive (i.e., intuitive) but primary emotional (i.e., instinctive). Hence, the brain structures involved may not be prefrontal but limbic.[28]

Consequently, it might be possible to have non-interpreted experiences. Hence, the RErl Absolute Unitary Being, might be experienced in the same

26 This will be returned to below.

27 Runehov (2004) 60.

28 Observe that the cognitive brain structures are present, thus the experiences are not to be confused with sensory impressions.

way that a newborn baby experiences things, i.e., the experiencer, at the very moment of the Absolute Unitary Being experience, does not interpret it primarily cognitively but primarily emotionally and directly in a newborn-baby-sense, i.e., instinctively. However, this would make the Absolute Unitary Being experience primarily limbic and this conclusion would not be accepted by all neuroscientists, for example, Azari, who correlates religious experiences with the cognitive structures of the brain instead of the limbic structures.

I believe that it has become clear by now that the cultural-religious-personal background of the experiencer is significant for his or her interpretation of the religious experience, because the cultural-religious-personal background of the experiencer will furnish his or her religious Erlebnis in a specific way in order to become a religious Erfahrung. Moreover, and most important for our discussion, the cultural-religious-personal background will even tailor the religious Erlebnis. The way of meditating, such as the via negativa, via positiva, transcendental meditation, or praying, is culture related. This cultural-religious-personal background of the experiencer may not only be significant for the interpretation of the religious experience by the experiencer but even for the religious experiences per se. In other words, not only will the Absolute Unitary Being Erfahrung be different but even the Absolute Unitary Being Erlebnis will. If this is right, it cannot be the case that all Absolute Unitary Being experiences are equal.

Assume that Marc practises the via negativa and Rita the via positiva meditation technique, and that both experience God or Ultimate Reality during one of their meditation sessions. Marc's experience will probably be an experience of wholeness, of unity, of absolute oneness, where there is no time, no space, no distinctions, no self. This is because, as we know, the via negativa is a meditation technique in which the meditator concentrates on emptying his or her mind completely. Rita's experience will probably also be an experience of union but of a union with the inner or outer image of her meditation, because, as we know, the via positiva is a meditation technique in which the meditator focuses on a chosen image.

Thus, not only will Marc and Rita describe their experiences in a different way (RErf) but they will also actually experience them differently (RErl) due to the different meditation techniques they use. The theoretical models Newberg and d'Aquili worked out illustrate this neurological difference. Nevertheless, they would maintain that, in the end, Rita and Marc will have the same experience, namely AUB, which might be the case or not.

Did Newberg and d'Aquili find the invariant element across the religions that they were looking for? The answer is that we cannot know. One problem is that neuroscience cannot demonstrate that there cannot be two different Absolute Unitary Being experiences. That there is a correlation with the total deafferentation of both PSPL (which was not confirmed by their neuroimaging experiment) is not sufficient, because of the impact that the cultural-religious-personal background of the experiencer may have on the experience per se (religious Erlebnis). Another problem is that even if we assume that their neuroscientific research has shown that there cannot be two different such experiences, why would the Absolute Unitary Being experience be the invariant element across all religions? Why not, among other possibilities, Einhorn's *concealed God* or Hick's *the Real*, or should we understand this concealed God and the Real equally? Some would even wonder whether there has to be a common core for all the religions. I believe that the idea that there cannot be two different Absolute Unitary Being experiences is challenged. It shall be challenged once more in the next section.

6. 4 Are All Experiences Equal?

The next conclusion drawn by Newberg and d'Aquili that we shall discuss is whether all experiences are equal. As we saw, they maintain that all human experiences are equal neuroscientifically and that everything we experience, we experience in a secondhand-way.[29] The idea is that all our experiences are mediated by our brain.

As mentioned in the analysis earlier, Newberg and d'Aquili maintain that if God exists and communicates with us so that we have a religious experience *x*, God cannot do this other than by using different neural pathways, *y*. They argue that neurologically, experiencing Absolute Unitary Being or God *x* and experiencing eating a piece of apple pie *x'* are identical concerning correlations with different neural pathways *y*. However, they do not maintain that *y* causes *x* or *x'*. What they argue is that the brain scans of a person eating a piece of apple pie *x'* show the neural activity involved in eating apple pie *y*'. For example, these scans show the olfactory areas that register the aroma, the visual areas picturing the piece of apple pie, the centres of touch holding the piece and the areas responsible for taste, in the same way that they show the neural activity *y* involved in experiencing Absolute Unitary Being *x*.

So far I am inclined to agree. However, Newberg and d'Aquili also maintain, as we have seen earlier, that "there can be little doubt that Absolute Unitary Being exists".[30] When they maintain that Absolute Unitary Being doubtlessly exists, they seem to want to give Absolute Unitary Being an identical reality status as a piece of apple pie, which will be further discussed below.

29 Newberg et al. (2001b) 36.
30 d'Aquili/Newberg (1996) 248.

Is it reasonable to claim that all experiences are identical on a neurological level? Consider the following thought experiment. Assume that Peter and Nancy eat a piece of the same apple pie. Will they describe their experience identically? Peter might say that the apple pie was not sweet enough while Nancy might say that it was too sweet. Peter might have been hungry when eating the pie while Nancy might have eaten it, because it was served at coffee time in the office. Does this not imply that different brain structures and functions had to be activated? For instance, the brain structure that has to do with being hungry will not be activated in Nancy's brain. I believe that Peter and Nancy's experiences of eating a piece of apple pie may not be equal neurologically. Similarly, if they had been practising the same meditation technique for twenty years and had been practising at least two hours per day I believe that they still would have had different religious experiences. The underlying principle is known as *the multiple realization principle* and dictates that even if we do the same things, we will still have different neural values.[31]

Another problem is that some neuroscientists claim that every brain is different and argue that brains are like faces. The idea is that there are no two identical faces, and not even identical twins have identical faces.[32] Similarly, there are no two identical brains. Furthermore, besides natural-aspects (genes, nutrition, disease, laws of physics and chemistry), we also have to consider nurture aspects (interaction with our environment (culture) from the day we are born until the end of our days), as well as personal aspects, i.e., our actions and choices. Every brain changes constantly and in an unique manner.[33]

People do things differently at different times and do different things which must imply that our brains develop differently and thus are different. For example, even if the identical twins, Maddy and Maggy, who were my classmates in elementary school, dressed alike and practically did the same things all the time, they did not do everything in the same order or with the same motivation or intensity. As identical twins, their brains might have looked very similar at the beginning of their lives but today, their brains are unquestionably very different.

Related to this is a last remark that every experience is associated with a complexity of neural activity in different brain structures. Carol Albright writes that the "many forms of religious experience may involve almost every part of the brain".[34] Thus, for an Absolute Unitary Being experience

31 We pray (alike) but what is going on in our brains differs.

32 I mean by nature. Advanced plastic surgery, today or in the future, might be able to create two identical faces. Another possibility for creating identical faces (and brains) might be by way of cloning techniques.

33 This is known as the plasticity of the brain.

34 Albright (2002) 167.

to occur, more neural structures are to be activated than the posterior superior parietal lobes. The PSPL may be important or central but by no means the only significant brain structure involved. This implies that if two experiences are to be identical, all the neural activity involved (and every neural process is a complex process in itself) must be identical.

Thus, I suggest that meditation and the accompanying religious experiences affect the Catholic nun and Rajiv's brains differently and that their RErl of Ultimate Reality may not be equal, not even neurologically. This suggestion also applies to all kinds of experience. What we can agree upon, however, is that all experiences are mediated through our brain.

Are all Absolute Unitary Being experiences equal then? The multiple realization principle, the natural and nurture aspects and the complexity of the brain, make it difficult to accept Newberg and d'Aquili's claim that all Absolute Unitary Being experiences are equal, even neurologically. It follows that, once again, their idea that Absolute Unitary Being could be the mutual core of all religions is challenged even if by *Absolute Unitary Being* they mean a *total deafferentation of both PSPL*, especially when we accept the assertion of neuroscientists that there are no two identical brains. In my view, it can be questioned whether there can be two identical total deafferentations of brain structures, for example, of the PSPL.

6.5 Does Absolute Unitary Being Exist?

Andrew Newberg and Eugene d'Aquili claim that Absolute Unitary Being exists.[35] Philosophically, this is a problematic assertion.

From a philosophical point of view they cannot maintain that God or Ultimate Reality exists. From the premises that (1) there is plenty of evidence in cross-cultural religious literature, (2) there are plenty of testimonies of living mystics and other people who experienced Ultimate Reality and (3) neuroscience experiments have confirmed the material basis for experiences of God it does not logically follow that God or Ultimate Reality really exists. That plenty has been written on experiences of Ultimate Reality does not, by itself, make God or Ultimate Reality real, nor, unfortunately, do the testimonies of people having had such experiences. Neither does the neurological basis of Ultimate Reality experiences by itself prove anything of the like, because what this neurological basis does prove is only that there is specific neural activity going on in the brain of the meditator at the moment when he or she maintains having his or her most intense religious experience. The conclusion does not follow from the premises.

35 d'Aquili/Newberg (1999) chap. 10, Newberg et al. (2001b) chap. 8.

Another problem we encounter is that from a purely neuroscientific perspective, it is not possible to maintain that Absolute Unitary Being, which they also refer to as God, exists or is absolute reality. As Peterson maintains, "[w]hile d'Aquili and Newberg's theory makes sense of the experiences of unity and bliss that mystics across the world have attested, it does not support the stronger religious claims regarding the origin and significance of these experiences".[36] Peterson argues that by referring to the descriptions of the mystics (the meditators) concerning what they experience, and by correlating these with the appropriate brain activity, Newberg and d'Aquili broaden the understanding of these experiences. However, this does not mean that Newberg and d'Aquili can explain the origin of the experience. The problem is, in my view, that they cannot detect any object of experiences by way of brain scans at all, nor can they detect any experience. The red, blue or green spots on the computer screen of the SPECT do not refer to an aspect of a subject or object of experience such as God or a piece of apple pie. What they refer to is neural activity. In other words, the blue or red or yellow spots that the neuroscientists see on the screen of the SPECT when scanning the brain of a meditator experiencing Absolute Unitary Being or eating apple pie are pictures of *neurochemistry* and not pictures of *God or pie*. Ilia Delio, for instance, writes in his response to Newberg et al. 2001, that "[i]f the researchers themselves based this claim on the data, [...] how [do they] proceed from biological structural changes to the reality of God"?[37] What he means is that neuroscientists, when engaging in neuroscience, lack an appropriate method to be able to draw ontological conclusions.

So how are we to understand Newberg and d'Aquili's idea that Absolute Unitary Being or God really exists? A first way in which we could understand their idea may be illustrated with the help of Saint Teresa of Avila who said:

> I do not say this as a comparison, because comparisons are never satisfactory: it is really true. The difference is similar to the one between something that is alive and something that is painted, no more nor less. Because, if I see a picture, it is a living picture – not a dead person but rather the living Christ. And he shows me that he is both man and God – not as he was in his sepulchral chamber, but as he was when he left it after that He was resurrected from the dead.[38]

Saint Teresa of Avila argues that she does not only interpret her mystical experiences. Rather, she maintains that it is really true and means that the object of the experiences which she describes in her interpretations really

36 Peterson (2002) 249.
37 Delio (2003) 578.
38 Peterson et al. (1999) 21. My translation.

exists. Now, remember that Newberg and d'Aquili maintain that a religious experience is real in the same way that when the experiencer recalls his or her experience, he or she remembers it in a similar way to the way that he or she remembers his or her non-religious experiences. They also maintained that the word real is used in the same sense as it is used when saying that the cherry tree in the garden is real.

If Newberg and d'Aquili understand the existence of God or Absolute Unitary Being in the way illustrated here, this would imply that they have to take Saint Teresa of Avila's experience literally. This, of course, need not be wrong. For instance, Saint Teresa's experience may have been a religious experience of the observable type. By observable is meant observable in the same sense that a cherry tree is observable.[39] This would imply that Newberg and d'Aquili have to understand Absolute Unitary Being or God as *ein Wesen* and not as *sein*, or that Absolute Unitary Being or God is able to appear before us as *an Ultimate Wesen*.

However, not everybody would agree with them and some would argue that Absolute Unitary Being or God cannot exist in the same way as the cherry tree, because to be able to experience, for example, the beauty of God, God would have to be at a certain distance from us to allow our sensory perception structures to transport the perception into other structures of the brain where it is transformed into an experience of God, in the same sense as into an experience of a cherry tree.

Nevertheless, Newberg and d'Aquili could argue that there is more to perception then space-time coordinates. Perception, for example, also involves seeing colours, hearing sounds, apprehending different smells, and involving, not least, feelings. For instance, assume that you see me embracing the man I love. There will be no doubt that the man I love exists; except everybody who is or has been in love knows that you need not be with him or her to know that he or she exists. Nevertheless, this understanding would meet the objection that Absolute Unitary Being or God cannot exist in the same way as the loved-one, since the beloved does not need to be physically present. This is because the loved-one always has the possibility of manifesting him- or herself in a four-dimensional reality, which it is said, Absolute Unitary Being or God does not. They argue that the loved-one can be observed by other people and thus be verified, which they maintain, is not the case for Absolute Unitary Being or God.

However, this arguement can only hold if the following type of religious experience, *the experience of God or Ultimate Reality with observable effect* were to be proved false. Otherwise, Absolute Unitary Being or God could be observed by more than one person in the same way as the loved-

39 The experience of God or Ultimate Reality with observable effects, see chapter 2.

one could be. For instance, take the example mentioned earlier of John seeing his friend Nathan being healed by Jesus touching his head. Nathan said that he *felt* the hand of Jesus, and therefore claimed that he was healed by him, while John said that he *saw* Jesus putting his hand on Nathan's forehead.[40]

A second way of understanding what Newberg and d'Aquili mean when they maintain that Absolute Unitary Being or God really exists is to understand the experience as an experience without an object. One has objectless experiences with absolute immediacy, i.e., one is aware of something but not through anything else. For example, I can experience a headache. I do not experience my headache through something else but I experience it immediately. My headache can become so intense that I am totally absorbed by it: my headache and I are indistinguishable. Then the experience has become an experience of absolute immediacy. I suggest that an Absolute Unitary Being experience or God may be a religious experience of absolute immediacy.[41] For instance, Alston maintains that

> [e]xtreme mystical experience in which all distinctions are transcended in an undifferentiated unity is properly thought of as absolute immediacy. If no distinction can be made within the seamless unity, then there is no possibility of distinguishing the experience involved from the object of awareness.[42]

However, these insights cannot be derived from neuroscientific investigations alone. Neither can neuroscientific explanations or other scientific explanations of God experiences, account for how, to put it in Gellman's words "God gets into the brain".[43] However, if Newberg and d'Aquili understand God or Absolute Unitary Being as the *absolute ground of everything that is and ever will be*, whether or not God is observable, accessible or understandable by science, philosophy or any other system of analysis, God or Absolute Unitary Being is already *in the brain* just as in everything else there is and Newberg and d'Aquili do not have to account for how God gets into the brain. However, if Newberg and d'Aquili consider Ultimate Reality or God as *Wesen*, the problem remains because as such God is not seen as the absolute ground of everything that is, but as the Creator of everything that is, distinct from everything that is.

Neuroscientists cannot maintain that Absolute Unitary Being exists because what is seen on the computer screen of the SPECT does not refer to God or Ultimate Reality but to neural activity. Nevertheless, correlating the

40 See further chapter two. God is understood as manifested in God's son Jesus.

41 We touched on this idea earlier when trying to reconstruct how newborn babies interpret their experiences.

42 Alston (1991) 22–23.

43 Gellman (2001) 101.

descriptions of the meditators and what they experience during their meditation session with the appropriate neural activity can broaden the understanding of these experiences but it cannot explain their origin. Still, there are questions that remain unanswered, such as how Newberg and d'Aquili understand God or Ultimate Reality. This has importance for understanding how *God gets into the brain*.

6.6 Different Thalamic Values

Newberg and d'Aquili's experimental study showed that there are differences in the brains of non-meditating people and the brains of the meditators, namely, the meditators had a significantly different thalamic laterality index at baseline compared to the control group of non-meditators.[44] Even if, as Newberg and d'Aquili maintain, more research is needed to be able to confirm or confute this result, several questions arise concerning the possible cause of these changes. These questions are:

Firstly, did these thalamic differences arise because of the many years of intense meditation, i.e., does frequent and intense meditation change the thalamic brain structures and functions? Sportsmen, for instance, develop bigger hearts than people who do not engage in sports on an elite level. Hence, why would the brain remain unchanged?

Secondly, were these meditators born with a physiologically different thalamus compared with the thalamus of the persons in the control groups? The latter question calls for another one, namely, how are these differences in the thalamus related to meditation and prayer? Do they predispose people to adopt meditation or prayer in their lives?

Thirdly, were the meditators born with certain talents, which during the years have developed into permanent changes of certain brain structures, for example, by way of living a monastic life? The idea is that the brains of the monks and the nuns are shaped to cope with this special lifestyle, including frequent and intensive praying and meditation, eating special food, fasting and deprivation of sleep, etc. These practices may lead to lasting changes in the brain.

The first option, i.e., that people who meditate change their brains in a significant sense is problematic, because we do not know how this change could be interpreted from a medical point of view. For instance, what consequences do these changes have for the health of these people? Furthermore, assume that the result of such a medical investigation was in favour of meditation, does this mean that meditation techniques should be taught in

44 Newberg et al. (2001a).

schools as a health improving subject similar to sports and gymnastics? Today, political groups discuss whether people who live a healthy life, i.e., do not smoke, are restrictive with alcohol, eat healthy food and/or engage in fitness activities, should receive better medical treatment. For instance, those persons who have a health-improving life-style would not have to queue in order to receive their medical treatment. If neuroscience proves the effects of meditation to be favourable for people's health, would this imply that practising meditation would be ranked among the health-improving factors that may guarantee people better medical treatment?

To agree upon the second option, i.e., that some people are born with specific features in their thalamus, might not only have consequences for contemporary neuroscientific theories but also for contemporary theological theories, as well as for the religions. This is because this could imply that not all people have the potential to have certain religious experiences, which could have the theological consequence that some could interpret this as *few are chosen* to reach Absolute Unitary Being or God. This really is a shocking thought.[45]

To agree upon the third possibility is, I believe, the most rational thing to do, because then, practising meditation becomes a matter of talent in the same sense that certain people have a talent for playing music, which develops under good conditions.

Newberg and d'Aquili have not developed possible consequences of their findings as I outlined above, but they raise the question whether "God creates the brain, or [...] the brain creates God"?[46] Is it possible to answer this question, and if so, how?

Consider the following thought. Could it be that as we are only able to create our everyday common reality based on the sensory input that enters our brain, we should also only be able to create Ultimate Reality based on the sensory input that enters our brain? Newberg and d'Aquili continue

> [w]hether we perceive something as aesthetically pleasant or not, whether we understand the reality of our experiences or not, whether we are closer to what is really "out there" or not, we ultimately create our own reality. We simply do the best we can at understanding ourselves and the world as we perceive them through our creative brain and creative mind.[47]

Thus, whatever we experience, and whether we understand our experiences or not, whether there is something outside our experiences or not, it is our

---

45 A similar discussion concerns the brains of geniuses. Do certain people become geniuses by way of adequate hard training or are they simply born and predestined to become geniuses, i.e., their brains being shaped differently to the brains of average people?

46 Newberg (2002) 145.

47 Newberg/d'Aquili (2000) 67.

brain and our mind that shape our world for us. Thus, in this sense, our brain creates God in the same manner that it creates the trees of the forest. This suggestion would be in agreement with Newberg and d'Aquili's claim that everything we experience, we experience in a second-hand way, i.e., through our brain (except for immediate awareness such as headache or the experiences of a new-born baby, perhaps).[48]

Another thought is, could it be that God creates the brain and that the brain creates God? I mean that God could have created us and that those of us who have a religious life create God (through prayer and meditation, in nature, in different events in life, etc). Newberg and d'Aquili write, "both conclusions about God (AUB) are in a profound and fundamental sense true, namely, that God is created by the world (the brain and the rest of the central nervous system) and that the world is created by God".[49] Thus, they suggest, that while human beings originate from God (God creates the brains), it is the brain that creates our concept of God. Intuitively one would like to agree upon this.

However, according to Mogens Dalby, neuroscientists, using merely neuroscientific means, are not able to show that the brain creates God. This is so because God cannot be read on the forehead of the person. Seen from the outside, the brain (brain scan) is not a medical book. Information from the outside of the brain has to be related to the information from the inside of the brain, i.e., from the person's descriptions. Only then a neurological diagnosis can be drawn.[50] An important remark is that the outside brain along with the inside brain create an information bank which is known as memory.[51]

Could it be that God creates the brain? This question cannot be answered by neuroscientific research either, because neuroscience, like all natural science, takes the material world as ontologically primary.

It thus seems that the only rational answer is that the brain creates God in the sense that all religious experiences are interpreted according to the experiencer's social-cultural-religious-personal frame of reference in the same way that the brain creates other human experiences. However, I want to emphasise that this consideration also means that God or Ultimate Reality may exist outside our experiences in a similar way to the forest.

---

48 Newberg et al. (2001b) 36.

49 d'Aquili/Newberg (1999) 193.

50 During our meeting in Århus, he explained that the brain has two sides, an inside and an outside. The inside brain is associated with a person's psychological and spiritual world. The outside brain is the information organ par excellence.

51 Note that memory is not only one single neural structure but is constituted from a complex system of brain structures and function.

6.7 Experiencing and Types of Experience

Once more it might be helpful to repeat what was said in chapter two. An experience (including a religious experience) *x* is an aspect of the subject of experience *y*. Furthermore, an experience consists of *y* and sometimes even of an object *y*', but not necessarily. Being thirsty would be an experience of an aspect of the subject, namely, the central nervous system, and the experience of drinking water would even have an object, namely water. We have also seen that there are several ways of (religious) experiencing. These ways are, (1) the *participating perceptive sensory way* of experience, (2) the *participating introspective sensory way* of experience, (3) the *participating introspective non-sensory way* of experience, (4) the *witnessing perceptive sensory way* of experience, (5) the *witnessing introspective sensory way* of experience and finally, (6) the *witnessing introspective non-sensory way* of experience.

It is important to keep in mind the following: (a) *Participating* here means to participate in one's (religious) experience in contrast to *witnessing* one's (religious) experience. It is the difference between participating in a mass and experiencing witnessing the mass on television. (b) *Introspective* refers to the (religious) experiencing of an aspect of a subject *y* in contrast to a *perceptive* which refers to the (religious) experiencing of *y* and an object *y*'. For instance, one may be experiencing reading the Bible in contrast to experiencing God's beauty in nature. (c) Finally, *sensory* means that something is experienced using the five senses rather than *non-sensory* which concerns thoughts, intuition and feelings. For example, one may experience a religious ceremony as opposed to experiencing Absolute Unitary Being.

Keeping this in mind, what are the ways in which the meditators experience? Let us first consider the meditators using the via positiva meditation technique in which they concentrate on an object somewhere in the room such as a cross on the wall. At the start of their meditation they will have a participating, perceptive and sensory way of experiencing. As mentioned earlier, to focus (on an image, a mantra, one's breath or on emptying one's mind completely) demands effort, which implies that one has to participate in the experience of focusing. The experience involves an object. The experience is of a perceptive kind and it is sensory because the meditator has to look at the cross on the wall. After a while though, the way of experiencing will change. When the meditator feels that he or she has become united with the cross, he or she will experience this in a participating, introspective, non-sensory way. This goes on, as Newberg and d'Aquili's neurological models suggest, until he or she *lets go*. He or she will then experience Absolute Unitary Being associated with a witnessing introspective non-sensory way of experiencing.

The meditators using the via positiva and focusing on an inner image and those who use the via negativa meditation technique (focusing on emptying their mind of all thought) will experience this focusing in a participating introspective non-sensory way. The experience is introspective because the experience concerns an aspect of a subject but not an object. It is non-sensory because focusing on an inner image or on emptying the mind of all thought does not involve any of the five senses. Then, when they reach Absolute Unitary Being, they will experience it in the witnessing introspective non-sensory way. People who use other types of meditation, such as, Transcendental Meditation, will probably experience this in a similar way, because all types of meditation contain both a moment of focusing and a moment of *letting go*. However, as we mentioned earlier, some meditators focus on a mantra i.e., a sound they produce themselves in order to put themselves in a meditative mood. I suggest that there are four ways of experiencing during meditation. The first is the participating perceptive sensory way of experiencing (outer image). The second is the participating introspective sensory way of experiencing (mantra). The third way of experiencing, then, is the participating introspective non-sensory way (inner image, focusing on one's breath or emptying one's mind of all thought, becoming one with the object of meditation and maintaining focus). Finally, we have the witnessing introspective non-sensory way of experiencing (Absolute Unitary Being).

To conclude, the way of experiencing will vary along the continuum, and furthermore along the process of meditation. The way of experiencing will, needless to say, also vary between the different meditation techniques (and/or between other religious practices.)

Which types of religious experience could be of concern to Newberg and d'Aquili's neuroscientific research on religious experience, obtained through meditation?[52]

I suggest that Swinburne's fourth type of religious experience, mediated by certain private sensations but not of the kind describable by normal vocabulary, is typically associated with the most intense moment of meditation, since the meditators often describe what they experience at the most intense moment of their meditation as ineffable.[53] For instance, Robert, one of the Tibetan monks says that

> [at the most intense moment of meditation] he suddenly understands that his inner self is not an isolated entity, but that he is inextricably connected to all of creation. Yet when he tries to put this intensely personal insight into words, he finds himself falling back on familiar clichés that have been employed for centuries to express the

---

52 Those religious experiences that may be ranked on the Aesthetic-religious continuum.

53 Swinburne (1979) 251.

elusive nature of spiritual experience. "There's a sense of timelessness and infinity, [...] [i]t feels like I am part of everyone and everything in existence."[54]

Furthermore, I believe that Swinburne's fifth type, experiencing God or Ultimate Reality not mediated by any sensation, is also relevant, because Ultimate Reality or God is often claimed to be experienced immediately as an immediate awareness, or intuitively. God or Absolute Unitary Being is experienced in the same way that one experiences one's own consciousness or feelings. For instance, one of the Franciscan nuns described her most intense moment of praying as

> a tangible sense of the closeness of God and a mingling with Him. [They would say:] How great is the mercy of the one who realized this union [...] I possessed God so fully that I was no longer in my previous customary state but was led to find a peace in which I was united with God and was content with everything.[55]

The descriptions of the Franciscan nuns are often inspired by (or similar to) those of the mystics of the past. This means that Newberg and d'Aquili may be right in that studying religious experiences obtained through meditation also allows them to understand spontaneously obtained religious experiences, such as Saint Paul's experience on his way to Damascus.[56]

Does this mean that the other four types of religious experience which were presented in chapter two are not relevant at all? Some of them are relevant. Those types of religious experience that might occur while meditating but which are not typically associated with the most intense moment of meditation are: (1) Swinburne's first type of experience of God or Ultimate Reality mediated through a common, public, sensory object.[57] For example, this might be to experience God's beauty in an icon or a sunset. A person may claim to have experienced God in that way while meditating on the beach (regardless of meditation technique he or she used). This is not the same as becoming one with the object of meditation though, as some persons describe their most intense moment during a via positiva meditation session. These meditators do not claim that they see God's beauty in the object, but that they feel one with the object. (2) His second type, experiences of God or Ultimate Reality through very unusual public objects is also relevant,[58] because the meditator might claim that she or he has witnessed an appearance, among other possibilities, of the Virgin Mary in the candle or in a cloud, while meditating or praying, again regardless of which meditation technique is used. (3) Another type of religious experience that

54 Newberg et al. (2001b) 2.
55 Newberg et al. (2001b) 7.
56 Acts 9: 1–22.
57 Swinburne (1979) 250.
58 Swinburne (1979) 250.

is relevant is Swinburne's third type which concerns experiencing God or Ultimate Reality mediated through private sensations that can be described in normal sensory language.[59] The meditator may have a vision. (4) The sixth type describing religious experiences which are observable by more than one person may, of course, also occur. However, they are not typically associated with meditation.

Whether these religious experiences can be studied by way of neuroimaging technology, I do not know. I suppose that it would not be impossible for the meditators to pull the twine every time they have a certain type of experience and that it would not be impossible for neuroscientists to scan what happens in the brain at these moments. The question is whether the information they received would make religious experience more understandable.

6.8 Summary and Conclusions

It is suggested that the neuroscientific experiment is scientifically valid but not optimally valid.[60] Even though Newberg and d'Aquili could have been more careful regarding the age and gender issues of the sample as well as the size of it, they were careful when they chose the type of religious experiences to study, since they restricted their research to the types of religious experience accessible to scientific research. The act of meditation can be performed in laboratory settings and is thus repeatable and testable. They seem to have chosen the most optimal equipment and method for their purpose, namely SPECT-imaging.

The remarks I gave on how Newberg and d'Aquili's neuroimaging study was performed were, firstly, they could have given the non-meditating participants a cognitive task in order to be able to compare the scans taken of the brains of the meditators and the non-meditators also at a later moment, instead of merely at baseline. Since meditation and prayer also correlate with the structures of the brain that have to do with cognition, they might have discovered interesting differences and/or similarities between the group performing a specific cognitive task and the group performing meditation. Secondly, they could have distinguished between the types of meditation when they took the images of the brains of the meditators. If they had, we might have been able to learn whether, and to what extent, their neurological models for religious experiences obtained by way of different meditation techniques correspond with the neuroimaging data.

59 Swinburne (1979) 251.
60 Newberg et al. (2001a).

The experimental study did unfortunately not confirm the models in the way that Newberg and d'Aquili may have hoped. However, several important correlations between the religious experiences and the brain were confirmed, if not completely, at least to some extent. Of special importance are the frontal lobes (cognition), the limbic system (emotions) and the PSPL (Absolute Unitary Being). They also discovered a difference in thalamic values at baseline between the meditators and the non-meditators. Thus, more neuroimaging studies (and similar) are desired.

Newberg and d'Aquili's idea that Absolute Unitary Being could be the invariant element across the religions has been challenged. One problem is that neuroscience cannot demonstrate that there cannot be two different Absolute Unitary Being experiences. That there is a correlation with the total deafferentation of both PSPL (which was not confirmed experimentally) is not sufficient because of the impact that the cultural-religious-personal background of the experiencer may have on the experience per se (religious Erlebnis). Another problem is that even if we assume that their neuroscientific research has shown that there cannot be two such different experiences, why would the Absolute Unitary Being experience be the invariant element across all religions? Some would even wonder whether there has to be a common core for all the religions.

The multiple realization principle, the natural and nurture aspects, and the complexity of the brain make it difficult to accept Newberg and d'Aquili's claim that all experiences are equal, even neurologically. In addition, there is the claim by neuroscientists that there are no two identical brains, which means that it could be questioned whether there could be two identical total deafferentations of brain structures. This implies, once again, that their idea that Absolute Unitary Being could be the mutual core of all religions is challenged, even if by Absolute Unitary Being they mean the total deafferentation of both PSPL.

Another important conclusion that was drawn was that neuroscientific results alone cannot prove that Absolute Unitary Being exists because what the neuroscientists see on the computer screen of the SPECT, when scanning the brain of a meditator experiencing Absolute Unitary Being, are pictures of neurochemistry and not pictures of God. Newberg and d'Aquili's studies may not be able to explain the existence of something simply because all clinical diagnoses need the narratives of the person that is diagnosed. However, correlating the descriptions of the meditators concerning what they experience during their meditation session with the appropriate neural activity can broaden the understanding of these experiences but it cannot explain their origin. Still, there are questions that remain unanswered such as how to understand God or Ultimate Reality. Newberg and d'Aquili equate Absolute Unitary Being with God but, unfortunately, it is not clear how Newberg and d'Aquili understand God, as Ultimate Sein or as an Ultimate Wesen.

It seems that the idea that the brain creates God is plausible. It means that all religious experiences are interpreted according to the experiencer's social-cultural-religious-personal frame of reference in the same way that the brain creates all human experiences. However, this consideration also means that God or Ultimate Reality might exist outside our experiences similarly to the forest. It was also said that if we have to choose between the presented alternatives of how practising meditation can be understood, it might be most rational to consider it as a matter of talent in the same sense that certain people have a talent for playing music.

Finally, the religious experiences that Newberg and d'Aquili were able to study were limited, both concerning the type of religious experience and concerning in what way the experiencers had them. It was said that there are four ways in which Newberg and d'Aquili's participants may have experienced during meditation. The first is the participating perceptive sensory way (outer image). The second is the participating introspective sensory way (mantra). The third way of experiencing, then, is the participating introspective non-sensory way (inner image, focusing on one's breath or emptying one's mind of all thought, becoming one with the object of meditation and maintaining focus). Finally, we have the witnessing introspective non-sensory way of experiencing (Absolute Unitary Being). It was concluded that the way of experiencing will vary along the continuum, and furthermore along the process of meditation. The way of experiencing will, needless to say, also vary between the different meditation techniques, (and/or between other religious practices.)

Only Swinburne's fourth and fifth types of religious experience seemed to be relevant for the experience during the most intense moment of the via negativa and via positiva meditation processes. Those types of religious experience were (1) the experience of God or Ultimate Reality mediated through private sensations that cannot be described in normal sensory language and (2) the experience of God or Ultimate Reality that is not mediated by any sensations. However, other types of religious experience may be relevant as well for other religious experiences ranked on the Aesthetic continuum.

Newberg and d'Aquili do not maintain that religious experiences are nothing but brain activity as Persinger does. They do not assert that religious experiences are illusory or that God or Ultimate Reality cannot exist. On the contrary, they seem to want to allow for the possibility that God or Ultimate Reality exists and they explain religious experiences in terms of correlation with specific neural activity. They also give authority to the testimonies of the meditators concerning what they experience during their meditation.

7 Neuroscientific Explanations of Religious Experience

We have finally reached the chapter in which I shall offer an answer to the core question of the present research project, namely, *in what way and to what extent neuroscientists can explain religious experience*. What exactly does this question mean?

Asking the question in what *way* neuroscientists can explain religious experience is asking by what *means* neuroscientists can explain religious experiences. But it is also asking *how* they can explain religious experiences, i.e., in terms of what. It is clear that the most important criterion for the neuroscientific study of religious experience is that the objects, phenomena or events of study are observable, or more exactly, that these experiences are measurable by neuroscientific means. These means may be, among other possibilities, ElectroEncephaloGram (EEG), functional Magnetic Resonance Imaging (fMRI), Positron Emission Tomography (PET), Single Photon Emission Computed Tomography SPECT or Magnetoencephalography (MEG). These neuroscientific measurement techniques measure neural activity in the brain (in different ways).[1] By measuring is meant to estimate or appraise something by a criterion. Such a criterion for neuroscience is a living healthy brain with a regular pattern of neural activity, deemed as a norm by neuroscientists. For instance, a PET scan of the brain of a person who suffers from schizophrenia will show a different level of neural activity in specific parts of the brain compared with a scan of the same parts of the brain of a healthy person (the criterion). To be able to perform such examinations, the person has to be prepared according to specific rules and prescriptions. In some cases the person has to perform specific tasks such as memory tasks, or has to sleep, depending on what the neuroscientists want to observe. The second criterion is that the study (experiment) is repeatable. Unique and unrepeatable events are thus not subjects for (neuro)scientific research. Finally, their explanations must be testable. This is also one of the reasons, as Newberg and d'Aquili maintain, why spontaneous religious experiences are very difficult to study by means of today's neuroscientific methods, because among other things, it is not

1 As mentioned earlier, the TMS that Persinger applies is not a tool of measurement, observation or diagnostics but a therapeutic and treatment tool for certain psychiatric disorders. Hence it does not measure but actually changes neural activity in the brain. We also saw that the application and its results are still discussed and questioned. For these reasons I do not take the TMS into account.

known which factors trigger spontaneous experiences, nor when such an experience may be expected. It is different for religious experiences obtained by way of meditation, rituals, prayer, and trance, since persistent training and endurance are known to be factors that trigger the religious experience. Furthermore, the exercises can be performed within laboratory environments and thus studied by neuroscientific methods.

How then do neuroscientists explain religious experience, i.e., in terms of what? Is there merely a correlation between certain religious experiences and certain neural activity or does certain neural activity even cause particular religious experiences?

Asking the question to what *extent* neuroscientists can explain religious experiences is also twofold. The question concerns which religious experiences can be studied: all of them, a limited amount, or none at all. The question also concerns whether neuroscientists can explain religious experiences they have studied exhaustively, partly, or not at all. The second question is subordinated to the first.

Evaluating the research done by Newberg and d'Aquili as well as by Persinger made it clear that the religious experiences that can be studied by today's neuroscientific methods are limited. Nonetheless, I could conclude that these religious experiences proved to be similar to all other experiences, i.e., aspects of a subject and sometimes, but not necessarily so, they even contain an object. Furthermore, religious experiences are those experiences in which it seems that God or Ultimate Reality appears to a person in a way that is in accordance with the doctrines and soteriology central to the religious traditions and their continuities. Newberg and d'Aquili studied religious experiences obtained by way of different meditation techniques. This allows the neuroscientists to study the following ways of religious experiencing: (1) The participating perceptive sensory way (focusing on an outer image). (2) The participating introspective sensory way (focusing on a mantra). (3) The participating introspective non-sensory way (focusing on an inner image, on one's breath or on emptying one's mind of all thought, becoming one with the object of meditation and keeping focused). (4) The witnessing introspective non-sensory way (Absolute Unitary Being experience). Following the possible ways of (religious) experiencing defined in chapter two, two ways were left out. The first is the witnessing perceptive sensory way of experiencing. For example, the experiencer would report that he or she has seen an angel in the same way that he or she usually sees his or her brother. The other is the witnessing introspective sensory kind which was described above as maybe being the most common religious experience, namely, seeing God in nature. Of course, it is not impossible that the meditators have religious experiences in these ways; however, such experiences are not typically associated with the meditation techniques mentioned above. These ways of experiencing are, instead, typically associated with spontaneous religious experiences which are more difficult, if not impossible, to study by neuroscientific means.

Concerning types of religious experience, we saw that only Swinburne's fourth and fifth types of religious experience seemed to be relevant for the experience during the most intense moment of the via negativa and via positiva meditation processes. Those types of religious experience were (1) the experience of God or Ultimate Reality that is mediated through private sensations "yet these are not a kind describable by normal vocabulary",[2] and (2) the experience of God or Ultimate Reality which the experiencer "does not have by having sensations".[3] However, other types of religious experience may be relevant as well. Hence, four were left out.[4]

To say that an explanation explains something to its full extent is to say that the explanation is exhaustive. Such an explanation then explains all the dimensions of that something in such a way that there is no need for any other explanation. We already know that neuroscience cannot explain all types of religious experience, for example, spontaneous religious experiences. Moreover, neuroscience alone is not able to explain non-spontaneous religious experiences in an exhaustive way either, because as mentioned earlier, the scientists need information from the experiencers (the meditators), concerning what they experience during the meditation process to be able to create neurological models for the different stages of experience that the meditators go through and to see how these experiences are related to the neural activity in different brain structures.[5] They also need information concerning the cultural-religious-personal background of the experiencers, such as the religion they belong to, which meditation technique they use, and so on. Hence, there are dimensions of religious experience that neuroscientific explanations alone cannot reach. This implies that neuroscientific explanations are necessarily limited explanations of religious experience. To what extent Newberg and d'Aquili as well as Persinger explain religious experiences will be further discussed below, starting by addressing reductionism.

Hence, the question was, in what way and to what extent can neuroscientists explain religious experience? The answer turned out as follows: In a way that today's neuroscientific means allow them to, that is, in terms of neural chemistry; and to the extent that religious experiences are observable

---

2 Swinburne (1979) 251.

3 Swinburne (1979) 251.

4 See evaluation chapter of Newberg and d'Aquili's neuroscientific research on religious experience above. They are also associated with spontaneous religious experiences which are more difficult to study (if possible) by neuroscientific means.

5 See Newberg and d'Aquili's models for the via negativa and the via positiva meditation techniques.

(measurable), repeatable and testable, and therefore, due to these criteria and to today's neuroscientific means, only partly.

Thus, neuroscientists can explain religious experiences in a methodologically restricted way and to a methodologically limited extent.[6]

7.1 Reductionism

The term *reduction* has different meanings depending on the context in which it is used. Reduction in science and recent philosophy of science is often understood as follows: Processes, properties, laws or theories found in higher levels of complexity can be entirely derived from those found in lower levels of complexity, ultimately in physics. John Dupré maintains that Paul Oppenheim and Hilary Putnam defend this kind of reduction because they emphasise a hierarchical classification of objects ordered in such a way that the objects at each level are composed exhaustively of entities from the following lower level. Their model further consists of what they call *bridged principles* in order to identify the objects at the higher level with specific structures of objects at the next lower level. Such a reduction program is transitive and will therefore end by showing that "the whole of science [derives] from nothing but the laws of the lowest level, or the physics of elementary particles, and the bridged principles".[7]

As mentioned earlier, since neuroscientific explanations or religious experiences are necessarily limited explanations, they are reductive in one way or another.[8] Thus, I believe it relevant to refer to some types of reductionism in order to elucidate exactly in what way and to what extent Newberg and d'Aquili, as well as Persinger, explain religious experiences.[9]

According to some philosophers, there are six established types of reductionism: ontological reductionism, reductive materialism, eliminating materialism, scientific reductionism, causal reductionism and methodological reductionism. However, by showing how these are related to one another I believe that these can be reduced to two types of reductionism, namely, ontological reductionism and methodological reductionism.

Ontological reductionism has the purpose of rendering an exhaustive explanation of some phenomenon, process or event (the *explanandum*) in

6 My answer, thus, is dynamic and not static.

7 Dupré (2000) 402.

8 In my opinion, all explanations are necessarily reductive in some way.

9 I shall not account for all of them, nor will I discuss reductionism as a whole. However, the topic is highly interesting and since I have some ideas worth further investigating, I shall consider them in my research project *Con-Science about Consciousness: The quest for a tenable Explanation of Higher State Consciousness Bridging Theology, Philosophy and Neuroscience*.

terms of another phenomenon, process or event (the *explanans*).[10] Accordingly, *something x* is explained in terms of *nothing but y*. A variant, for example, is to maintain that everything that exists is made up of a small number of basic substances that behave in regular ways like atoms, quarks or energy. Then one has applied ontological reductionism. Nancey Murphy understands ontological reductionism as "no new kinds of metaphysical 'ingredients' need to be added to produce higher-level entities from lower level ones".[11] Her definition rejects the existence of vital forces or entelechy in the life sciences, as well as mind or soul as the basis of consciousness. She also designates a type of reductionism that has even stronger claims than ontological reductionism, namely, *reductive materialism* which dictates that all higher-level explananda can be exhaustively explained in terms of lower level explanantia with the addition that only the lower level explanantia are really real while the higher-level explananda are only composite structures made of atoms.[12] The latter type of reductionism makes the ontological reductive arguement, that religious experiences are nothing but experiences caused by God, impossible, because God is apprehended as a higher-level explanans, which cannot be reduced to composite structures made of atoms. Still, in my view, reductive materialism is a form of ontological reductionism, even the strongest form.

Yet another type of ontological reductionism, and which is gaining ground, is known as *eliminating materialism*, which dictates that mental states such as religiosity and love are simply misnomers. One day all such concepts will be seen as inadequate and it will become clear that what is at stake is pure neural-chemistry. Eliminating materialism projects the success of science in explaining misnomers in the past on to the future. The exhaustiveness hence lies in the future.[13] Which form of ontological reductionism is applied depends on how the explanans is defined. When the explanans is a scientific explanation, we are dealing with *scientific reductionism* which is a form of ontological reductionism when it is argued that phenomena, processes and events can exhaustively be explained by science. It is the same *causal reductionism* when it is intended that the explanandum can be exhaustively explained by what caused it.[14] Thus, *reductive materialism* is

---

10 The definition of ontological reductionism derives from Holte (1984) 45. Mikael Stenmark defines ontological reductionism in a similar way as Ragnald Holte, see Stenmark (2001b) 21–22. The terms explanandum and explanans are taken from the DN model presented in chapter two. However, here the explanandum will intend the phenomena, process or event to be explained and the explanans will stand for the phenomenon, process or event offered to account for the explanations of the explanandum.

11 Murphy (1998b) 129. Murphy defends non-reductive physicalism.

12 Murphy (1998a) 466–467/(1996) 65.

13 As we have seen, Patricia Churchland utilizes such a reductionism.

14 Causal reductionism is actually understood as a bottom-up hierarchy from, which if true, it

an extreme form of ontological reductionism while *scientific reductionism* and *causal reductionism* may be forms of ontological reductionism if the claim is that the explanans exhaustively explains the explanandum. What all types of ontological explanations have in common is that it is the explanans that exhaustively explains the explanandum or at least will be able to do so in the future.

A different type of reductionism is known as *methodological reductionism* which can also be understood in different ways. Firstly, it can be understood as every explanation ought to be continually reduced to the very simplest possible explanation. Secondly, it may be apprehended as a research strategy for analysing things to study, such as cells, in terms of their parts, such as, macromolecules, as well as for applying successful theories in one area, such as Darwinian evolution, to other areas, such as sociology or religion. Thirdly, methodological reductionism may aim to reduce a whole into a specific set of parts for a scientific purpose without claiming that this is all there is to that whole.[15] An illustration of such reductionism would be to explain pain in terms of its neurological correlations. For instance, such a neurological explanation would be an example of a scientifically reductionist explanation that is not a form of ontological reductionism, because the scientific explanans is not meant to exhaustively explain the explanandum (pain is not maintained to be nothing but neural activity).

To round off, I suggest there are two main types of reductionism: ontological reductionism and methodological reductionism. The difference between these two types of reductionism is that ontological reductionism signifies that phenomena, processes or events can be exhaustively explained by other phenomena, processes or events. Methodological reductionism, on the other hand, means that a whole is reduced to a specific set of parts for a scientific purpose without claiming that this is all there is to that whole. Hence there is no claim for exhaustiveness.

7.2 Accurately Reducing

The present question is how reductionism may be applied. In my view, reductionism can be utilized accurately or erroneously.

An accurate way to reduce a phenomenon, process or event is such that the explanans highlights some element, some important aspect(s) belonging to the explanandum while there is no claim that the explanans exhaustively explains the explanandum. To reduce something in an adequate manner then

---

would follow that the laws of the sciences ranked higher up in the hierarchy should be reducible to the laws of physics.

15 See further, Stenmark (2001b) 21–22, Holte (1984) 44–45.

is to partly explain something. As such love could be explained in terms of hormonal activity but it would not be claimed that there is nothing more to love. The idea is that the explanandum can be explained in different ways depending on different perspectives. The explanans may be uninformative but not irrelevant, i.e., a phenomenon may be explained in terms of some part of the phenomena that is not informative for the time being, but to explain it in terms of something irrelevant would be wrong. To illustrate what is meant, assume that a mathematician provides a reductive explanation of some astronomical behaviour by way of a specific equation. That equation will be uninformative for all who are not acquainted with higher level mathematics, but the explanation is nevertheless adequate. On the other hand, explaining why the poinsettia has red petals by saying that it is a beautiful plant which in North European countries is appreciated most during the Christmas season, would be irrelevant and inadequate, because this explanation has nothing to do with the question of why the poinsettia has scarlet petals.

From what is maintained in the previous section, we are able to establish that methodological reductionism has to be an accurate form of reductionism. This is because, by way of methodological reductionism, one explains or elucidates some part or aspect of a certain explanandum from a specific point of view. This point of view may, for example, derive from scientists, but also from theologians, who, using a specific scientific method, explain a part of the explanandum of study in accordance with what they are able to (being methodologically restricted), or want to (within the realm of the restrictions), explain using that specific method.

Let us continue with some examples. A first example comes from Rudolf Otto. He maintains that religious experiences (he means mystical experiences) cannot be explained, that they are inexpressible because they are numinous (i.e., mystical). However, he does not mean that they cannot be explained at all, only that they cannot be explained in normal language.[16] Otto says that

> [the experience] grips or stirs the human mind. The feeling of it may at times come sweeping like a gentle tide, pervading the mind with a tranquil mood of deepest worship. It may pass over into a more set and lasting attitude of the soul, continuing, as it were, thrillingly vibrant and resonant, until at last it dies away and the soul resumes its "profane," non-religious mood of everyday experience. It may burst in sudden eruption up from the depths of the soul with spasms and convulsions, or lead to the strongest excitements, to intoxicated frenzy, to transport, and to ecstasy. It has its wild and demonic forms and can sink to an almost grisly horror and shuddering.[17]

16 Mystical experiences belong to the second type of religious experience, namely, the experience of God or Ultimate Reality mediated through an unusual, public, sensory object mentioned earlier.
17 Otto (1970) 12–13.

To Otto, the experience is ineffable. As I see it by asserting this, Otto explains a religious experience in terms of its awesomeness, overpoweringness, of energy and urgency, of the Wholly Other and of Fascination.[18] Otto explains religious experiences in a methodologically reducing religious way. More precisely, he explains them by theological means, highlighting a specific part of religious experience without claiming that his explanation is exhaustive. One could perhaps say that he applies a reductionism similar to scientific methodological reductionism.

Otto's way of explaining religious experiences is not different from other theologians who try to explain them in terms of specific aspects or elements of religious experience. Thus a second example of accurately reducing religious experience comes from Friedrich Schleiermacher. He does not explain religious experience in terms of its non-natural aspects but in terms of feelings. Schleiermacher views religious experience as "'immediacy of awareness' [...] of 'being utterly dependent upon God'".[19] He explains religious experience as an immediate, eternal, mystical moment when a person becomes aware (immediacy of awareness) of the all-surrounding, invasive, incredible presence of God in and through the universe (absolute dependency).

Schleiermacher defines the feeling of absolute dependency as an immediate consciousness of the relationship between Ultimate Reality or God and oneself in a way that is absolute and indistinguishable. He understands *feeling* similarly to how he understands intuition, namely, as an immediate cognitive relation to some object giving an insight that is not mediated by any concepts. Schleiermacher's definition of intuition is a modification of Kant's, that intuition is that through which a mode of knowledge is in immediate relation to objects. Schleiermacher uses the term *feeling* to describe religious experience in *On Religion*,[20] and absolute dependency in *The Christian Faith*.[21] Thus, Schleiermacher's *feeling* is not the kind of feeling, for example, as feeling hungry, but as feeling that such and such is the case. Schleiermacher especially addresses the fifth type of religious experience, namely, experiences of God or Ultimate Reality that are not mediated by any sensation. Schleiermacher also explains religious experiences by theological means and thus seems to embrace methodological reductionism.

The third example of accurately reducing religious experiences is in the way that William James explains religious experience. William James also explains religious experiences in terms of intuition and writes that

18 Otto (1970) 13/19/23/25/31.
19 Schleiermacher (1976) 26–27, as cited by Thiselton (2002) 277.
20 Schleiermacher (1799)/(1923).
21 Schleiermacher (1821)/(1976).

> [y]our whole subconscious life, your impulses, your faiths, your needs, your divination, have prepared the premises of which your consciousness now feels the weight of the result; and something in you absolutely knows that that result must be truer than any logic-chopping rationalistic talk, however clever, that may contradict it.[22]

We recognise the sudden and absolute knowledge similar to Schleiermacher's notion of intuition. However, James refers to the noetic (cognitive) quality or aspect of religious experience by saying that the experiencer remembers his religious experience as being immediate rather than the result of conscious thought and inference. To put it in Nina Azari's words, a cognitive process that nonetheless feels immediate.[23] James emphasises that the experiencers describe their religious (mystical) experiences as more real than any ordinary perception: "The something seemed close to me, and intensely more real than any ordinary perception. Everything else might be a dream, but not that."[24]

He explains:

> One may indeed be entirely without [religious experiences]; probably more than one of you here present is without them in any marked degree; but if you do have them, and have them at all strongly, the probability is that you cannot help regarding them as genuine perceptions of truth, as revelations of a kind of reality which no adverse argument however unanswerable by you in words can expel from your belief.[25]

The experience, according to James, is real in the same sense as the experience of being in love. Even if one's attention is addressed to other matters one cannot forget the loved-one, he or she "uninterruptedly affects [one] through and through".[26] James not only addresses the fifth type of experience but, I would argue, all six types. James, similar to Otto and Schleiermacher, utilizes methodological reductionism. I suggest the form of reductionism to be causal reductionism and that the causal factor then is a reality suddenly revealing itself to the experiencer. This explanation, however, is not exhaustive, because James also emphasises the relationship between the experiencer and the experienced. Thus, here we may have an example of causal reductionism as a form of methodological reductionism.

A final example of accurately reducing religious experiences given by theologians that I shall present considers how Jonathan Edwards explains them, namely, in terms of a new sense of the heart. In *Treatise on Religious Affections*, he writes that genuine religious affections consist in a new sense of the heart which is analogous to our five ordinary senses apart from that it

---

22 James (1997) 74.
23 Azari (2001) 1649.
24 James (1997) 64.
25 James (1997) 73.
26 James (1997) 73.

cannot be reduced to one of them. The new sense of the heart can only be grasped by experience. Edwards writes: “Affections that are truly spiritual and gracious, do arise from those influences and operations on the heart, which are *spiritual, supernatural and divine*”.[27] Thus, Edwards explains religious experiences in terms of direct and specific sense perception. This sense perception affects the heart, which in Western religious traditions is often associated with feelings deriving from God or Ultimate Reality. Edwards addresses thereby all six types of religious experience. In my view, his explanation also falls within the scope of methodological reductionism.

These explanations of religious experience are accurate reductionism, because they do not include exhaustive claims, i.e., religious experiences are not explained in terms of *nothing but*. Rather, they highlight specific and important aspects of religious experiences. The explanations of Otto, Schleiermacher, James and Edwards are methodologically reductionist, because they reduce a whole (religious experience) into a specific set, part or aspect (feelings, noetic quality, etc.) for a specific (theological) purpose without claiming that this is all there is to say about religious experiences.

In my opinion, Newberg and d’Aquili explain religious experiences in a similar methodologically reductive way when, for example, they explain Absolute Unitary Being in terms of its correlations with a total deafferentation of both the right and left posterior superior parietal lobe (PSPL). Let us return to Newberg and d’Aquili’s experimental study, without repeating the whole process (that Robert, one of the meditators, sits in a lotus position, holds a twine which he has to pull at the most intense moment of his experience, and so on.)

Assume that Robert pulls the twine at the most intense moment of his experience, i.e., when he experiences Absolute Unitary Being. The moment of Robert’s experiencing Absolute Unitary Being will be depicted by the scans of his brain as a total deafferentation of both the right and left PSPL. I want to remind you once more that the experimental study only confirmed the theoretical model to a certain extent. However, let us keep it for the sake of arguement. Newberg and d’Aquili will report this in the research report of that experiment. Assume that they have been able to repeat the experiment and could report the same result, i.e., Absolute Unitary Being experiences correlate with a total deafferentation of both the left and right PSPL.

Neuroscientifically, they describe religious experiences (of that kind) in terms of that specific neural condition or state. They methodologically

27 Edwards (1959)/(2004) part III.

reduce the experience to neural activity for the purpose of finding its neural correlations and of understanding the phenomenon better. They do not, however, reduce religious experiences to nothing but neural activity.[28] They do not maintain that there is nothing more to an Absolute Unitary Being experience than the correlating deafferentation of both the left and right PSPL. Newberg and d'Aquili maintain that the fact that they are able to picture the neural activity that correlates with a person experiencing eating an apple pie does not mean that eating an apple pie is merely brain activity. According to them, this is similar for Absolute Unitary Being experiences. It does not mean that because they are able to picture the correlating neural activity that Absolute Unitary Being experiences are merely neural activity. However, they claim that neuroscience has important information to add to the overall understanding of religious experience in the form of the different brain structures and functions that correlate with different religious experiences or specific neural activity that correlates with several religious experiences and what that means neurologically.

Thus, this way of explaining religious experiences, even if it is reductive, is not erroneous in my view, because it leaves open the possibility of comparing and eventually developing different (reduced) explanations (more on this later). Furthermore, it is not erroneous because this kind of reduction may actually capture important (or new) elements that belong to the experience and which are valuable for the overall understanding of religious experience. Examples from what has been said above are feeling of union; feeling of total dependency; feeling of insight; sense of the heart; the deafferentation of the PSPL.

The point I would like to make is that different aspects of a religious experience are elucidated depending on whose view is being accounted for, for example a theologian's, a neuroscientist's, a believer's or an experiencer's point of view. Religious experiences are explained to a restricted extent, i.e., to the extent that a certain religious experience correlates with a certain part of the brain, a certain feeling, a certain way of experiencing or a certain insight and what that means.

We may conclude that Newberg and d'Aquili explain religious experiences in the way that neuroscientific means allow them to and what that means neurologically, being aware of the shortcomings of (neuro)science. Furthermore, they explain religious experiences, limited to those obtained through meditation, in terms of which neural activity *y* (the explanantia) correlates with different religious experiences *x* (the explananda) and to the extent that *y* (the explanantia) correlates with *x* (the explananda) but *y* (the explanantia) cannot exhaustively explain *x* (the explananda).

---

28 Which would be ontological reductionism.

Newberg and d'Aquili, as well as Otto, Schleiermacher, James and Edwards explain religious experiences according to methodological reductionism, i.e., religious experiences are reduced into a specific set, part or aspect (neural activity, emotions, etc.), for a specific purpose, but it is not claimed that this is all there is to religious experiences.

7.4 Erroneously Reducing

To reduce something erroneously is to explain something exhaustively in terms of *something different* (the roses you saw were not roses but stones), in terms of *nothing but* (love is nothing but hormonal activity, i.e., there is nothing more to love than that) and in terms of an illusory or imaginary status (your headaches are but your imaginations, you cannot have a headache). In other words, the explanans exhaustively explains the explanandum. The type of reductionism under consideration is ontological reductionism. Why is that so?

The following example of what an ontological reductionist explanation could look like may be a good illustration of why ontological reductionism is wrong.

Ontological reductionism dictates that music is nothing but sound waves, i.e., air that vibrates in different sequences and with different amplitudes. However, we know that music can only occur when these sound waves meet a listener who interprets the sounds and comprehends them, according to his or her cultural, biological, and personal background, as harmonies and disharmonies. The person's mood will be an influencing factor for whether he or she experiences the sounds as music, as enjoyable etc. We also know that music has a source, one or many instruments, one or many voices or both. One and the same sound may be experienced as noise or as music, for example, the sound of a helicopter in Pink Floyd's *The Wall*.

I believe it is clear that ontological reductionism cannot be correct when essential aspects of an explanandum are left out. Ontological reductionism only seems to *capture the fish that the fisher's net is able to hold. The question is what happens to the other fish?*[29] To explain music exhaustively in terms of sound waves is wrong because there are other dimensions to music than the physical sound waves. These other dimensions are ignored by ontological explanation.

A clear example of ontological reductionism is the way in which Burrhus Frederic Skinner explains mental processes in terms of behaviour. However, his comment is that he does not reduce anything at all because there has never been anything to reduce: there are no mental processes. Thus,

29 Author unknown.

consciously or not, Skinner applies reductive materialism. To him, mental processes are not real but behaviour is.

In my opinion, Persinger explains religious experiences in such an erroneous reductive manner. His conclusions are such that neuroscience has shown the real nature of religious experiences and that this real nature has nothing to do with some Ultimate Reality or God. In other words, *x* explains the occurrence of *y* and is the real nature of *y*. Looking more closely at Persinger's explanations of religious experiences, it becomes clear that he applies ontological reductionism, more precisely, he applies scientific reductionism in an ontological reductionist manner. More precisely, he explains religious experiences exhaustively by way of scientific explanations. It may be interesting to look more closely at how he proceeds.

One way in which Persinger adopts scientific reductionism is by describing the explanandum in a way that fails to identify an emotion, a practice or an experience in the way the subject identifies it. An example of such reductionism would be to describe an experience in non-religious terms when the experiencer describes it in religious terms. According to Proudfoot, this is to misidentify the experience, or to attend to another experience altogether. Proudfoot's remark is justified. For instance, Persinger does not explain religious experiences. Rather he explains Epileptic seizure experiences, i.e., he attends to another experience altogether.

To illustrate what Proudfoot means by misidentifying an experience, imagine that Preben is afraid of bears. If Preben describes his fear of bears to a psychologist and the psychologist answers him that there are no bears in Copenhagen and what frightened him was probably the shadow of a tree, the psychologist would misidentify Preben's fear. It may very well have been the shadow of a tree that frightened Preben the last time he was wandering in the parcs but that is not the point. Nor is it that there are no bears in Copenhagen. Preben is afraid of bears, no matter whether there are bears in Copenhagen or not. Similarly, assume that Preben describes his experience of God to the psychologist (Persinger) and the psychologist answers him that there are no gods and what he experienced must have been an hallucination, then the psychologist has misidentified Preben's experience of God. Preben may have been mistaken in his experience but that is not the point. The point is that Preben believes in God. Even if Preben were to agree that he was mistaken this time, this would probably not stop him from believing in God and believing that the previous religious experiences he had were real.

Persinger also uses ontological scientific reductionism when he exhaustively explains religious experiences in scientific terms in such a way that the experiencer could not possibly approve of. (In the example above, the experiencer might agree with the given explanation.) Proudfoot maintains, however, that this way of reducing is perfectly justifiable and is, in fact, normal procedure.[30] What has to be explained, the explanandum, is set in a new context. Even if this way of explaining things and events is a normal procedure (within science) and that in some cases it is also justified, the way in which Persinger explains religious experience is not. A good example would be when he explains religious experiences in terms of subconscious memories of the parents or parents-substitutes.

Remember the story where Veronica was about to drown but was saved, according to her by God as an answer to her prayer. *God led* her to a floating log.[31] Persinger, however, asserted that what saved her was a simple log that *coincidentally* was floating in the water, and puts her experience in another context. He maintains that what made her think it was God who came to her rescue was that she, subconsciously, associated God, who saved her, with her parents saving her from different dangers when she was a little girl. Veronica, who prayed and cried for God to come to her rescue, would certainly not agree with Persinger's explanation. However, as mentioned earlier, this is not to exclude the possibility that there may be cases in which the experiencer mistook her/his experience.

Persinger's explanation, I maintain, is ontological reductionism and an erroneous reduction of Veronica's religious experience and other such religious experiences, that he considers are entirely the results of an adult reliving his or her infancy in a modified way. This way of explaining religious experience is not justified.

A second illustration of Persinger explaining religious experience in such an explanatory reductive way is when he explains religious experiences in terms of evolution, when he maintains that God Experiences (the experiences we erroneously take as having God or Ultimate Reality as their cause) are possible, because the human brain is shaped with the capacity for God Experiences, since such experiences were necessary for the survival of the human species. Religious experiences are a product of how the human brain responds to danger and nothing else. Once again, according to Persinger, science has exhaustively explained religious experiences.

Thus far, Persinger explains religious experience exhaustively in terms of something different, which the experiencer denies completely or which

---

30 I can agree that it may be a normal procedure when it comes to methodological reductionism but not when ontological reductionism is utilized. Proudfoot makes a distinction between identification and explanation. However, following his thought would lead me to an interesting but for my purpose unnecessary sidetrack.

31 As mentioned earlier, whether this example concerns a matter of miracle or not is not essential. What is essential is how she explains her experience (she does not say that it is a miracle but that God answered her prayer) and how Persinger puts her explanation, the explanandum, in another context.

he or she, even if he or she does not agree, might eventually accept as another plausible explanation. In the latter case, the experiencer may accept the explanation simply because it comes from a scientist and the experiencer may not have the opportunity to test the validity of the explanation. However, Persinger goes even further in his reduction of religious experiences. In his overall explanation of religious experiences he applies, in my view, reductive materialism. He says:

1. Religious experiences are illusory because they can be exhaustively explained by (neuro)science.
2. God is nothing but words (dog spelled backwards).
3. *Therefore*, all religious talk can be eliminated.[32]

Accordingly, Persinger not only applies ontological reductionism, but also reductive materialism and actually means that religious experiences are illusory and the only thing that is real is the explanans of religious experiences, such as, the neural activity of the brain or the underlying biology of personal crisis.

The reason Persinger and other neuroscientists think this way may be a consequence of what neuroscientists have been able to discover during the last decades, not least because of the development of imaging techniques such as fMRI, PET and SPECT. For instance, one of the latest discoveries by way of PET imaging technique is the discovery of where exactly in the brain the amyloid plaque, which is implicated in Alzheimer's Disease progression, is formed. The reason this is a special discovery and interesting for our debate is because until now this amyloid plaque has been the "*unseen enemy*".[33]

Because neuroscientists are now able to discover exactly in which parts of the brain the amyloid plaque *is formed*, which is key to tracking Alzheimer's Disease progression, by way of PET imaging technology, some would conclude that they can discover in exactly which parts of the brain a religious experience *is formed* by the same or similar imaging techniques. The *unseen* then becomes God instead of the plaque.

We may conclude that Persinger explains (what he refers to as) religious experiences in the way neuroscientific means allow him to do.[34] However, he does not seem to be aware of the shortcomings of (neuro)science. He explains religious experiences *x* (the explananda) in terms of which neural

32 Dewi Zephaniah Phillips calls this way of reducing unconscious reductionism. However, I shall keep to the types of reductionism described above.

33 Klunk (2003).

34 Except when his research is biased and particularly tailored.

activity *y* generates religious experiences (the explanantia) and to the extent that *the explanantia* exhaustively explains *the explananda*. In addition, he maintains that only the (relevant) neural activity *y* (the explanantia) is real while the religious experiences *x* (the explananda) are illusory. Persinger applies ontological reductionism, more specifically, scientific reductionism, and argues that science (by way of different explanans) can exhaustively explain religious experiences along with reductive materialism, i.e., only the explanans is real while the explanandum is illusory. He describes religious experiences exhaustively in scientific terms and fails to identify the experiencer's own descriptions of what they experienced, or he ignores the religiosity of the experiences completely but explains religious experiences exhaustively in terms of something completely different. It may be that Persinger's explanations are a consequence of jumping to the unwarranted conclusion that because neuroscientists are able to solve one problem by way of a certain scientific technique or method, neuroscientists are able to solve everything that they intend to study. Nevertheless, Persinger's explanations of religious experience are controversial and highly speculative and, as seen earlier, often contain badly documented claims. His conclusions are often too hastily drawn, sometimes even based on biased experiments.

7.4 Combining Explanations

In the previous section it was maintained that explanations of religious experiences are reductive explanations. However, different reductive explanations may be combined. Some explanations may exclude, enrich or weaken one another, and other explanations may complement one another etc. Let me briefly present some examples.

As shown above, Newberg and d'Aquili's neuroscientific explanations of religious experiences support the idea that all religions have a common core. The invariant element across the religions, they maintain, is to be found in the Absolute Unitary Being experience, which according to them, all meditators, regardless of which meditation technique they use and regardless of which religions they belong to, experience as the absolute final point of meditation. In my opinion, Newberg and d'Aquili's pluralistic view on the religions can be combined with and thereby enrich Hick's pluralist hypothesis, (and vice versa). The invariant element across the religions Hick calls the Real, an essentially unknowable object in itself, which manifested by way of religious experiences, is interpreted differently in accordance with people's cultural-religious-personal background. The Real is neither personal nor impersonal but becomes personal or impersonal when interpreted by its experiencers.[35] Thus, at first sight, the two pluralistic views could be combined and enrich one another. The question is only, exactly how would these two ideas enrich one another?

What will enrich Hick's pluralist hypothesis is Newberg and d'Aquili's hypotheses that Absolute Unitary Being is equal regardless of which meditation technique used and, that (according to Newberg and d'Aquili) the meditators describe what they experienced (RErf) during their Religious Erlebnis (RErl) Absolute Unitary Being in different ways. Hence, there is only one RErl, namely AUB, but there are different RErfs of that AUB. The meditators who belong to an Eastern religion will probably describe their most significant experience as an experience of wholeness, of unity, of absolute oneness, where there is no time, no space, no distinctions, no self. The meditators who belong to an Abrahamic religion will probably describe their most significant experience as an experience of unity with the God of their religion (God may be experienced as Absolute Unitary Being but also as, for example, an (inner or outer) image of Jesus). Thus, following what is said in the previous chapters on the work of Newberg and d'Aquili, an experience of (union with) God as Absolute Unitary Being should correlate with the total deafferentation of both the left and right PSPL, while the experience of (union with) God in the image of Jesus should correlate with a total deafferentation of the left PSPL and a partial deafferentation of the right PSPL.[36]

However, the combination of the two pluralistic explanations is not entirely successful. According to Hick, everyone who experiences the Real interprets his or her experience according to his or her cultural-religious-personal background. These backgrounds settle the different neural correlations and pathways. Newberg and d'Aquili also maintain that everyone who experiences Absolute Unitary Being interprets his or her experience according to his or her cultural-religious-personal background. However, an important difference with Hick's view is that to Newberg and d'Aquili, it is one and the same neural correlation that settles the experience of Absolute Unitary Being, namely, the total deafferentation of both the left and right Posterior Superior Parietal Lobes.

Thus, Newberg and d'Aquili's and Hick's pluralistic views are not interchangeable but there are elements that enrich one another's explanations.

We also saw that Einhorn suggests a concealed God (being neither personal nor impersonal) as the invariant element across the religions. Einhorn's idea is that this concealed God is found in all religions and becomes personal or impersonal due to different cultural-religious-personal interpre-

35 Hick (1989) 246–249.

36 Until the experiencer stops concentrating on the image of Jesus.

tations. The question thus is whether or not Newberg and d'Aquili's (and Hick's) pluralistic views can be combined with and would enrich Einhorn's. Einhorn, Newberg and d'Aquili and Hick all refer to an invariant element across the religions and therefore their pluralistic views can be combined. Even if Einhorn does not explicitely mention that the concealed God is manifested by way of religious experiences, Newberg and d'Aquili's and Hick's views could be a complement to his, and thus their idea's could enrich one another. However, the fact that the pluralistic views of Hick and Einhorn as well as of Newberg and d'Aquili could complement and enrich one another does not mean that all the problems concerning such a view are solved. One such problem concerns the truth claims that different religions make regarding Ultimate Reality or God.[37]

A view that would exclude Newberg and d'Aquili's pluralistic views is called *exclusivism*. Exclusivists maintain that "salvation, liberation, human fulfilment, or whatever else one considers as the ultimate goal of the religion, is found solely in or through one particular religion".[38] Exclusivists do not reject the idea that other religions may contain truths, but only one is exclusively effective and able to provide true salvation. For instance, the theologian Karl Barth asserts that "in the event of Jesus Christ, God uniquely provides and reveals the means for our reconciliation with God".[39] The idea is that without the revelation of God or Ultimate Reality, religions are inadequate, and such Revelation is only possible through Jesus Christ. Thus, the view that God reveals God-self in accordance with the doctrines and soteriology of the different religions (and their continuities) is wrong, because God only reveals the path to salvation in the event of Jesus Christ, making Christianity the only true religion. Hence, Barth's exclusivist view excludes Newberg and d'Aquili's as well as Hick's and Einhorn's pluralistic ones.

The last view I want to present weakens pluralistic explanations. This view is called *inclusivism*. Similar to exlusivists, inclusivists also defend one religion, but have in common with pluralists that they hold that God or Ultimate Reality can be encountered in various ways and through different religions. The theologian Karl Rahner adopts an inclusivist position and maintains that Christianity is the only true religion but God is love and wants everybody to be saved and hence God will reveal God-self even in other religions.[40]

---

37 Peterson et al. (2003) 276. Since pluralism, inclusivism and exclusivism, though interesting, are not main topics of the present research project, I shall not evaluate the problems related to these views.

38 Peterson et al. (2003) 270.

39 Peterson et al. (2003) 270–271.

40 Peterson et al. (2003) 280–281.

7.5 Explanatory Models

Since neuroscientists can only explain religious experiences in a methodologically restricted way and to a methodologically limited extent, neuroscientific explanations of religious experiences have to be reductive in one or another way. However, other explanations of religious experiences were also shown to be reductive. An important question becomes how different such reductive explanations may be related. In this section we will explore the kind of models that Persinger and Newberg and d'Aquili might use to explain religious experiences.

Newberg and d'Aquili as well as Azari explain religious experiences obtained by way of meditation in terms of correlations with specific brain activity.[41] Newberg and d'Aquili explain them in a methodologically reductive manner since they do not claim that their explanations are exhaustive. Since at least some of their views are also combinable with others, their explanatory model (or way of explaining) as a whole, is open for extension.

I associate extending with the manner in which railroad networks extend. An amount of different railroad tracks can build a railroad junction: Similarly, a number of different explanations can build a junction to extend the comprehension of, among other possibilities, religious experience (a specific one or not) or a specific religious phenomenon.

Newberg and d'Aquili's, Hick's, and Einhorn's ideas of an invariant element across the religions could build such a junction extending the comprehension of the pluralistic view of religions. Moreover, like one railroad track joins another, one explanation may join another. Such is the case for Newberg and d'Aquili's and Azari's neurological explanations of religious experience. Such explanations may conflict in part (Newberg and d'Aquili's and Azari's do) and one explanation could even be excluded when found less plausible. Furthermore, just as two railroad tracks remain parallel, never crossing and never being part of a junction, two explanations can remain aloof from one another and be regarded by the explainers as equivalent explanations. For example, some would understand the relationship between the big bang theory and the theory of creation as an explanation of the origin of the universe as such. Others would regard the relationship as conflicting or partly conflicting but compatible.

I argue that all of the combinations contribute to the extended overall comprehension of, in our case, religious experience.

41 Azari does not agree with me that the subjects of her experiment performed a kind of meditation. I believe this because they were reading a specific psalm over and over again, (and were brain scanned) and later they were memorizing this psalm and repeating it in thought over and over again (and were brain scanned again). According to me, both tasks but especially the latter, could be seen as if the subjects were repeating a mantra, which is a meditation technique.

Persinger's explanatory model is different, contradictory even. Because of his ontological reductionist way of explaining religious experiences, he fails to recognise other possible explanations (such as those given by the experiencers themselves). He does not have regard for scientific explanations that show outcomes different from his own (those recognising or supporting the reliability of religious experience and those doubting the power of science to be able to confirm or confute that reliability). Hence, even if Persinger's and Newberg and d'Aquili explanations exclude one another (religious experience being illusory versus real), Persinger's explanations could not join those of Newberg and d'Aquili's in the sense suggested in the extending explanatory model. Persinger's explanatory model, in my view, *absorbs*, as it were, other possible explanations. Therefore, we should not classify his explanatory model as an extending one. Persinger, I believe, would sympathize with Freud's way of explaining religious experiences since Freud considers religious experiences to be nothing but expressions of subconscious suppressed feelings.[42] Hence, Freud's explanation of religious experiences could be included in (absorbed by) Persinger's view that religious experiences can be exhaustively explained by (neuro) psychology. Let me suggest that we call Persinger's model a *totally absorbing explanatory model*.[43]

7.6 Developing Newberg and d'Aquili's Model

Since the totally absorbing explanatory model of religious explanations does not seem to be encouraging for further development, I shall concentrate on the extending explanatory model as applied by, among others, Newberg and d'Aquili and investigate its possibilities. Newberg and d'Aquili not only explain religious experiences but they also explore them, because their model not only explains what is already known but also gives us new knowledge and insight. Therefore, I shall, from now on, call their model the extending *exploratory* model.

How could the extending exploratory model of Newberg and d'Aquili be further developed? The possibility I have in mind has its origin in Wesley Wildman and Leslie Brothers' attempt to explain one kind of religious ex-

42 However, I am somewhat doubtful whether Persinger would recognise the existence of a subconsiousness.

43 Some would argue that there are also theologians who explain religious experiences by way of the total absorbing explanatory model. They think of those theologians who maintain that religious experience cannot be explained. However, to say that religious experiences cannot be explained need not imply that all other possible explanations are absorbed. Rather it may signify that we cannot explain religious experience at all, by any means. Such explanatory models simply *exclude* any *possibility* of explaining religious experiences. Therefore they are not of relevance for our purpose.

perience called *experience of Ultimacy*. An experience of Ultimacy is equal to Newberg and d'Aquili's Absolute Unitary Being experience or an experience of God or Ultimate Reality.[44] Could their idea be helpful in broadening Newberg and d'Aquili's explanation of Absolute Unitary Being and/or vice versa? How do Wildman and Brothers proceed?

To begin with, Wildman and Brothers try to explain the experience of Ultimacy as objectively as possible by combining neuroscientific, psychological, sociological, theological and ethical insights, and secondly, based on this, they provide procedures for identifying typical and authentic experiences of Ultimacy and specify the causes and values.[45] Both efforts are interesting, but to continue our discussion it will suffice to look more closely at the former one, i.e., how they describe Ultimate Reality by way of combining explanations from different disciplines. However, where their quest is to find an explanation which is as objective as possible, my goal is rather to illustrate a way of finding a tenable explanation of religious experiences based on different explanations of religious experiences including neuroscientific ones.[46]

Wildman and Brothers include four perspectives in their explanation of Ultimacy experiences, two of which derive from phenomenological and social-psychological understandings. The other two components derive from neurology and theology. In this way they collect *markers* which together allow for a broad understanding of experiences of Ultimacy. These markers are: Phenomenological characteristics; neurological signatures; the descriptions of the people within their own social-linguistic contexts; the judgement of experts in religious discernment or of psychologists; and the wisdom of generations encoded in theological and ethical traditions. By running a reported religious experience through this entire collection of markers, they hope to determine whether this experience actually is an experience of Ultimacy.

In the first group we find, for example, Otto's phenomenology of numinous experiences, Kierkegaard's three-staged phenomenology of religious conversion, and other theological-phenomenological descriptions. The second group contains neuroscientific information such as the correlations between religious experiences and the brain. According to Wildman and Brothers, neurological correlates may be of use to distinguish false from real Ultimacy experiences in spite of what the experiencer maintains. If this

44 Wildman/Brothers (1999).

45 Wildman/Brothers remain neutral as to the reality or non-reality of the experiences and of what is taken for the object of Ultimacy experiences. However, they do realise their limits in this exploit.

46 To search for a tenable explanatory model for explaining religious phenomena is a goal I have set up for my present research project.

were possible, would this imply that only those experiences that correlate with a total deafferentation of both the left and right posterior superior parietal lobes (PSPL) could be considered as experiences of Absolute Unitary Being, God or Ultimate Reality? Let us investigate this.

Picture ten meditators who participate in an experimental study of religious experience performed by Newberg. Assume that the study is scientifically valid. Assume further that all ten meditators practise the via negativa and that their task is to meditate and to pull the twine when they feel that they experience Absolute Unitary Being (they know exactly what it is like to have such an experience). One of Newberg's hypotheses is that there cannot be two different Absolute Unitary Being experiences because of the correlation with a total deafferentation of both the left and right PSPL. Assume that all ten meditators pull the twine at a certain moment and that all of them describe their experience completely in agreement with what it is like to have an Absolute Unitary Being experience. However, when he compares the pictures of their brains taken at the moment of the Absolute Unitary Being experience, Newberg discovers that eight show a total deafferentation of the left and right PSPL but two do not. One picture shows limbic activity instead and the other only a tendency towards a deafferentation. How should these results be understood, following Wildman and Brothers?

I believe it depends on what information we receive from the other markers. (The reported religious experiences have to be run through the entire collection of markers, not only the neuroscientific one.) However, possible explanations why some of the meditators' experiences did not correlate as was predicted could be that (1) both experiences were faked; (2) only one was faked, the other meditator just pulled the twine a little too early and therefore there was only a tendency towards the total deafferentation; (3) none were faked, one meditator pulled the twine too early and the other had misunderstood the instructions and pulled the twine when he or she had an ecstatic and blissful feeling which correlated with a spillover in the hypothalamus' vertromedial portion which maximally stimulated the ergotropis sympathetic system, i.e., limbic neural activity, (4) the hypothesis that there cannot be two different Absolute Unitary Being experiences has to be reconsidered. Hence, even if neurological correlates may be useful in some way to distinguish false from real Ultimacy experiences in spite of what the experiencer maintains, we cannot substantiate that they will succeed in their task.[47]

47 Furthermore, our discussion in the evaluation chapter above puts this idea a little bit in the shade. Rather, our discussion is in agreement with Wildman/Brothers' other hypothesis that Ultimacy experiences are widely distributed across the brain, which does not confute that neuro-

The third group consists of information coming from psychology and sociology. Psychologists then contribute with descriptions of such *typical* experiences.[48] Sociologists who understand the influences of religious groups on people will be able to contribute with information concerning the complex social interactions within which Ultimacy experiences often occur. The fourth and last group entails theological theories of Ultimacy. Among others, they present the theory that Ultimacy experiences transform people's character, identity, life or lifestyle.

The advantage of Wildman and Brothers' model is that even though the challenge of reductionism remains unchanged, it avoids hasty reductive conclusions by considering relevant data from different sources. However, in my view, their model does not sufficiently consider the experiencers' descriptions of what they experience during an Ultimacy experience. Their focus lies mostly on the information coming from several and different *experts*.

Some would argue that since psychologists participate in the team of experts and since their task is to account for the subjective aspects of the experiences, an additional description from the experiencers themselves is superfluous. I agree that psychologists are trained to study the subjective aspects of a person's experiences but they do this by way of psychological means. The results are not always in agreement with the interpretations of the experiencers themselves. However, assuming that the psychological explanations are not ontologically reductive, they may complement, enrich, weaken or even exclude the experiencers' narratives. This is why, in my opinion, both are needed.

Newberg and d'Aquili's exploratory model not only considers the descriptions of the experiencers but they also give these descriptions authority. One example is their taking seriously the experiencer's description of an Absolute Unitary Being experience as more real than everyday reality. Furthermore, from these descriptions, along with the correlating neuroscientific markers, Newberg and d'Aquili conclude that Absolute Unitary Being exists. On the other hand, as we discussed in the evaluation chapter above, they do not pay enough attention to *expert* information. I suggest that their exploratory model would benefit if they were to include information concerning the cultural-religious-personal backgrounds of the experiencers given by, it is suggested, sociologists, theologians, philosophers of religion and psychologists.

---

scientific criteria of authenticity will be used to determine whether a religious experience is real or not, but it confutes that neuroscientific criteria would be a deciding factor.

48 Wildman/Brothers (1999) 354.

7.7 Summary and Conclusions

The core question focused on in my research, i.e., in what way and to what extent neuroscientists can explain religious experiences, has been addressed. The answer was, in a way that today's neuroscientific means allows them to and thus in terms of neural chemistry, and to the extent that religious experiences are observable (measurable), repeatable and testable and, due to these criteria and to today's neuroscientific means, only partly. In other words, neuroscientists can explain religious experiences in a methodologically restricted way and to a methodologically limited extent. We saw that Newberg and d'Aquili explain religious experiences in the way that neuroscientific means allow them to, being aware of the shortcomings of (neuro)science. Furthermore, it became clear that they explain religious experiences, limited to those obtained through meditation, in terms of which neural activity correlates with different religious experiences and to the extent that neural activity (the explanantia) correlates with religious experiences (the explananda) but neural activity (the explanantia) cannot explain religious experiences (the explananda). On the other hand, Persinger explains religious experiences in the way that neuroscientific means allow him to but he does not seem to be aware of the shortcomings of (neuro)science. He explains religious experiences in terms of which neural activity (the explanantia) generates religious experiences (the explananda) to the extent that religious experiences (the explananda) can exhaustively be explained by (neuro)scientific explanations (the explanantia). Furthermore, the explanantia (the scientific explanations) are real while religious experiences (the explananda) are illusory.

To be able to answer the core question I have, in addition to what was examined in the previous chapters considered some types of reductionism. In order to determine to what extent Newberg and d'Aquili and Persinger explain religious experience, I have investigated which types of reductionism Newberg and d'Aquili and Persinger applied when they explain religious experience. It was concluded that Newberg and d'Aquili explain religious experiences according to methodological reductionism. This means that religious experiences are reduced into neural activity for a specific purpose, but that Newberg and d'Aquili do not claim that their explanations exhaustively explain religious experiences; on the contrary. Persinger, it was shown, applies ontological reductionism. In other words, he maintains that science can fully explain religious experiences. Furthermore, we saw that Persinger also applies reductive materialism because he maintains that religious experiences are illusory, completely reducible to neural activity, which is what is real. I could also conclude that Newberg and d'Aquili reduce religious experiences in an accurate way while Persinger reduces them in an erroneous manner. Similar to music being explained exhaustively in terms of sound waves, which was said to be erroneous since there are other dimensions to music than the physical sound waves and these other dimensions are ignored or disregarded by ontological explanations, so Persinger's ontological explanations disregard other dimensions belonging to religious experiences.

It was shown that different explanations of religious experiences can be combined in order to broaden the comprehension of religious phenomenon. These explanations then complete, enrich, weaken or exclude others. By investigating Newberg and d'Aquili's and Persinger's way of explaining, I could designate two explanatory model, namely, the extending explanatory model (used by Newberg and d'Aquili) and the totally absorbing explanatory model (used by Persinger). Since Persinger's explanatory model is not encouraging for further development, only the extending exploratory model was considered due to its potential for further development. If Newberg and d'Aquili extend their model (which includes not only neuroscientific explanations of religious experiences obtained by way of different meditation techniques but also the descriptions of the experiencers of what they experienced), by adding to it the expertise of, for example, sociologists, theologians, philosophers of religion and psychologists, they would have an exploratory model that even if it adopts methodological reduction, will avoid hasty reductive conclusions. In my view such a model could be fruitful for broadening the understanding of religious experiences, even if the religious experiences that can currently be studied by neuroscience are limited.

8 Final Conclusions

The question of whether religious experiences could be experiences of God or Ultimate Reality or whether religious experiences are merely a product of our nervous system has been the focus of my research assignment. In other words, could religious experiences be *sacred* or merely *neural*? The starting point for my philosophical analysis has been that neuroscientists place different explanations at our disposal concerning what religious experiences are and what causes these experiences. The aims were (1) to clarify which religious experiences neuroscientists are able to measure by way of today's neuroscientific methods, (2) to analyse and to critically evaluate Persinger's and Newberg and d'Aquili's research that has been performed on religious experiences, (3) to clarify the way in which and to what extent neuroscientists can explain religious experience and (4) to find possible ways of developing tenable explanatory models for religious experience. The core question I decided to ask was *in what way and to what extent can neuroscientists explain religious experience?* The analysis and the evaluation were limited to a selection of research on religious experience done by Andrew Newberg and Eugene d'Aquili and Michael Persinger even if the work of other neuroscientists has helped me to perform the evaluation. To reach my goal, I identified Newberg and d'Aquili's and Persinger's neuroscientific explanations and their rationales, clarified what is meant by religious experiences and by explaining something and I also designated terms and definitions. The time has now come to constitute a resumé and to draw some final conclusions. It will become clear how the aims were fulfiled.

Are religious experiences sacred or neural? The result of my study shows that religious experiences are associated with both the sacred and the neural in the following sense. A religious experience is sacred when the experiencer recognises his or her experience as an experience of God or Ultimate Reality in such a way that it can be classified as such in accordance with the doctrines and soteriology central to the religious traditions and their continuities (usually, but not necessarily, the religious tradition to which the experiencer belongs). A religious experience is neural in the same sense that all human experiences are, i.e., every religious experience (every human experience) correlates with a particular brain activity. Accordingly, the answer to the question "Sacred *or* Neural"? turned out to be "Sacred *and* Neural".

The core question of my research project was in what way and to what extent neuroscientists can explain religious experiences. The answer became: In a way that today's neuroscientific means allow them to and thus in terms of neural chemistry; and to the extent that religious experiences are observable (measurable), repeatable and testable and, due to these criteria and to today's neuroscientific means, only partly. Hence, it was maintained that neuroscientists can explain religious experiences in a methodologically restricted way and to a methodologically limited extent. However, neuroscientists may or may not agree. Newberg and d'Aquili do. Persinger, however, does not. Rather, he believes that he explained religious experience to an exhaustive extent. Even if I only considered selections of the studies performed on religious experiences by two neuroscientific research teams, in my opinion, the result of my study could still be regarded as a basis for analysing and evaluating the statements of other neuroscientists concerning religious experience. This foundation consists of the following issues.

Firstly, it consists of an effective method by which to perform a philosophical analysis and evaluation, namely, a conceptual and argumentative philosophical method. Hence, we have a specific language consisting of relevant terms and concepts for interpretation of the neuroscientific research that we want to analyse and evaluate.

Secondly, this foundation consists of a definition of a human experience, i.e., an experience is an aspect of the subject of the experience and sometimes, but not necessarily, of an object. Since a religious experience, similar to a non-religious experience, is an interpretation of a human experience, the definition also counts for religious experiences. The aspects of the subject of experience and the objects, then, concern God or Ultimate Reality. However, even if these conditions are sufficient for having a human experience, for an experience to be a religious one, at least the following criteria have to be fulfiled: Firstly, it has to be an experience in which it seems that God or Ultimate Reality, being the absolute ground of everything that is (and ever will be), appears to a person in a way that is in accordance with the doctrines and soteriology essential to the religious traditions and their continuities. Secondly, the experiencer has to recognise the experience as a religious one, i.e., the experience has to be effective evidence and not merely de facto evidence. God or Ultimate Reality was thus understood as the absolute ground of everything that is (and ever will be). Having defined religious experiences in such a way gives us the advantage of being able to quickly recognise whether or not it is religious experience that the neuroscientists are interested in studying. For instance, we could detect that Newberg and d'Aquili actually studied religious experiences while Persinger rather studied epileptic seizure experiences, schizophrenic hallucinations or sensed presence experiences obtained by way of Transcranial Magnetic

Stimulation, which is a therapeutic treatment tool for certain psychiatric disorders and not a measurement instrument. Since I have also designated six possible ways of religious experiencing, presented six different types of religious experience and clarified the similarities and differences between religious and psychical experiences, it is also possible to evaluate the religious experiences studied by the neuroscientists in more detail. Hence, we were able to see that religious experiences obtained through meditation were experienced in at least four different ways. Concerning the types of religious experience, two of the defined types seemed to have been relevant. Having defined God or Ultimate Reality in a specific way helps us evaluate the conclusions drawn. For instance, we were able to see that Newberg and d'Aquili do not always understand Absolute Unitary Being in a consistent manner. Sometimes it was understood as Absolute Unitary Sein and sometimes also as an Absolute Unitary Wesen. Thus the first aim was achieved, i.e., to clarify which religious experiences neuroscientists are able to measure by neuroscientific means.

Thirdly, the foundation contains models for how religious experiences can be explained. Both explanatory and exploratory models can be used depending on whether something merely has to be made intelligible, in which case an explanatory model will be created, or whether something novel can be added to the explanation, in which case an exploratory model will be designed. This knowledge will be helpful for evaluating in which way neuroscientists explain religious issues similar to the way in which I was able to see that Newberg and d'Aquili used an exploratory model to reach their goal while Persinger used an explanatory model for his purpose. Furthermore, it will help to construct adequate models for our own investigations. Furthermore, since different ways of explaining things and events were highlighted and the differences and the similarities between them were considered, we have a tool to recognise whether neuroscientists explain religious issues in a scientific manner only or whether they blend different explanatory facilities.

Fourthly, by carefully analysing and evaluating the research on religious experience performed by Newberg and d'Aquili and Persinger, the foundation can be extended with descriptions of erroneous and accurate ways of reduction. An erroneous reduction of some religious explanandum is such that it can be exhaustively explained by a non-religious explanans. Additionally, such an erroneous reduction may include the claim that only the explanans is real while the explanandum is illusory. Persinger reduced religious experiences in such a way when he maintained that because neuroscience can explain religious experiences exhaustively, religious experiences are illusory and God is nothing but words. Religious experiences, for example, are explained in terms of epileptic seizure experiences or other occasional or lasting brain or psychological disorders which cause the temporal lobes to become abnormally active. The point is always that religious experiences are illusory, and that there is nothing sacred about them.

An accurate reduction of that same religious phenomenon (the explanandum) would be such that an aspect of that explanandum is explained by a specific explanans but this explanans cannot entirely explain the explanandum. Methodological reductionism was said to be an accurate way of reduction. Newberg and d'Aquili explain religious experiences in a methodologically reductive manner, because even if they explain them in a neuroscientific way, using the means that neuroscience can put at their disposal today, they recognise the shortcomings of their discipline. Thus, scientific (and also causal) reductionism may be ontological or methodological depending on whether the claim is that the explanans exhaustively explains the explanandum or not. Newberg and d'Aquili's explanations of religious experiences are in terms of which brain activity correlates with different religious experiences. They do not claim that there is nothing more to say about religious experiences; on the contrary. Needless to say, this knowledge helps us to recognise the way and the extent to which neuroscientists explain the religious phenomena they have studied. Hence, the second and third aims were fulfiled, i.e., to analyse and to critically evaluate the research performed on religious experiences done by Persinger and Newberg and d'Aquili and to clarify in what way and to what extent neuroscientists can explain religious experiences.

Fifthly, the foundation contains two explanatory models that I designated: (1) The extending explanatory model which has the advantage of being able to build explanatory junctions extending the comprehension of the religious phenomena that is studied, to remain apart from other possible explanations and to be able to join or refute other explanations. Such a model was used by Newberg and d'Aquili. (2) The totally absorbing explanatory model which does not recognise any rival explanations, which was used by Persinger. Designating those two models will make it easier to determine whether or not a neuroscientific explanation invites further development.

Finally, on the basis of the extending exploratory model as used by Newberg and d'Aquili, the following was suggested. Explanations deriving from different experts concerning religious issues should be added to the explanations given by the religious persons concerned. In doing so we would avoid hasty reductive conclusions but would instead be able to create conditions for constructing tenable exploratory models for religious issues. Mapping religious experiences, even if the type of religious experience is limited, demands the cooperation of different disciplines. Such a research team could perhaps consist of theologians; scholars of religious studies such as psychologists of religion; sociologists of religion; also philosophers of religion; and neuroscientists, especially neuroradiologists. I consider the fourth and last aim to be achieved, i.e., to find a possible way of developing a tenable explanatory model for religious experiences. As I mentioned in the introduction, it is also my expectation that my work will show that it is important that the philosophical, theological and neuroscientific disciplines are perceived as having vital contributions to make to one another. As such it is meant to be a contribution to the academic discipline of science and theology.

8.1 Some Suggestions to Neuroscientists

It should be clear by now that I appreciate Newberg and d'Aquili's neuroscientific research on religious experience. More generally speaking, I believe that neuroscience has important data to contribute to the overall understanding of religious experiences. Today the religious experiences that are of interest for neuroscientific research are limited to those which can be obtained by way of different meditation techniques and other cognition-based techniques such as reading and memorizing religious texts, or rituals. This is because these religious practices can be performed in laboratory settings and the studies can be repeated and tested in the way science prescribes. Until today, neuroscientists have had no possibilities of capturing and measuring spontaneous religious experiences like those similar to Saint Paul's revelation on the road to Damascus. Even in theoretical studies, spontaneous religious experiences are much more complicated to use as a basis for a theoretical neuroscientific model of the correlations between the brain and the experiences than those obtained through cognitive tasks such as meditation.

Along with the evaluation on the scientific validity of Newberg and d'Aquili's experimental study above, I had two aspirations. I would like to put these forward as suggestions to Newberg and other neuroscientists who want to study religious experiences by way of neuroimaging technology. The first suggestion is that a cognitive task would also be given to the non-meditating control group. For example, the participants could be asked to concentrate on a poem (repeatedly reading or rehearsing it). The advantage would be that neuroscientists could compare the scans taken of the brain of the meditators and the non-meditators at different moments, baseline and later on. Hence, even the scans taken at the moment that the meditator pulls the twine and those taken at the moment that the non-meditator ends the cognitive task according to the instruction that he or she had received could be compared, instead of only those scans of the brains of both groups at baseline.

The second suggestion concerns whether and to what extent the neuroscientific explanatory models for religious experiences obtained by way of different meditation techniques correspond to the results of the experimental study. Do the pictures of the brain taken by way of neuroimaging technology on a person performing the via negativa meditation technique correspond to the exploratory model of which parts of the brain were expected to have increased or decreased neural activity when meditating with this kind of meditation technique? Similarly, do the pictures of the brain taken by way of neuroimaging technology on a person performing the via positiva, numinosa, and TM, meditation technique correspond to the model of which parts of the brain were expected to have increased or decreased neural activity when meditating with these kinds of meditation techniques? It would be interesting to see whether or not these four ways of religious experiencing, connected to the meditation techniques, correspond to four different pathways. To obtain such a result I suggest that neuroscientists not only pay attention to the neural activity that is specifically active during the meditation process, but also to the type of meditation technique that is performed at the moment of the scanning.

Combining these neuroscientific results with the cultural-religious-personal information concerning the life and history of the meditating persons gained by the team of experts suggested above and the descriptions of the meditators themselves concerning what they experienced during the meditation sessions, would give us a more comprehensive understanding of religious experiences obtained by way of different meditation techniques. We would understand better what the meditators experience when, among other things, they experience bliss, religious awe, Ultimate Reality, etc. We would learn how they arrive at those experiences; what happens in the brain; what consequences these experiences have in their lives; how their characters changed after the first experience and how it further develops; why they use the meditation technique they use; why they started meditating; whether they belong to a certain religion and if so which religion they belong to; and so on. Let me end this study by citing one of Albert Einstein's thoughts concerning religious experiences:

> The most beautiful and profound emotion we can experience is the sensation of the mystical. He to whom this emotion is a stranger, who can no longer wonder and stand rapt in awe, is as good as dead. To know that what is impenetrable to us really exists, manifesting itself as the highest wisdom and the most radiant beauty, which our dull faculties can apprehend only in their primitive forms – this knowledge, this feeling, is at the center of true religion.

Albert Einstein

Bibliography

Albright, C.R. (2001) "Neuroscience in Pursuit of the Holy: Mysticism, the Brain, and Ultimate Reality", *Zygon*, vol. 36 (September) 485–492.

Albright, C.R. (2002) "Religious Experience, Complexity & the Image of God", *Neurotheology, Brain, Science, Spirituality, Religious Experience*, 167–182.

Alston, W.P. (1991) Perceiving God: The Epistemology of Religious Experience, Cornell University Press: New York, USA.

Arts, N. (2001) *Epilepsy Through the Ages: An Anthology of Classic Writings on Epilepsy*, Van Zuiden Communications: Alpen aan den Rijn, The Netherlands.

Ashbrook, J.B. (1993) "From Biogenetic Structuralism to Mature Contemplation to Prophetic Consciousness", *Zygon*, vol. 28 (June) 231–249.

Ashbrook, J.B./Albright, C.R. (1997) *The Humanizing Brain*, Pilgrim Press: Cleveland, USA.

Atran, S. (2002) "The Neuropsychology of Religion", *Neurotheology, Brain, Science, Spirituality, Religious Experience*, University Press: San José, USA, 147–166.

Audi, R. (1999) *The Cambridge Dictionary of Philosophy*, Cambridge University Press: Cambridge, UK.

Azari, N. (2001) "Short Communication: Neural correlations of religious experience", *European Journal of Neuroscience*, vol. 13, 1649–1652.

Avila, St.T. (1995) *The way of Perfection*, Peers, A.E. (ed.), Scanned by Harry Plantinga.

Baldwin, J.D./Baldwin, J.I. (1998) *Behavior Principles in Everyday Life*, Prentice Hall: New Jersey, USA.

Barbour, I.G. (1990) *Religion in an Age of Science*, The Gifford Lectures (1989–1991) vol. 1, Harper/Row: New York, USA.

Barbour, I.G. (1996) *Issues in Science and Religion*, Harper/Row: New York, USA.

Barendregt, H. (2001) "Het moeilijke probleem, Relatie tussen de hersenen en bewustzijn", *Hersenen, Bewustzijn, Zich op onszelf*, Oomen, P. (ed.), Valkhof Pers: Nijmegen, The Netherlands.

Beckmann, A. (1992) "Supervenience, Emergence, and Reduction", *Emergence or Reduction? Essays on the Prospects of Nonreductive Physicalism*, Beckmann, A./Flohr, H./Kim, J. (ed.), Walter de Gruyter: New York, USA, 94–118.

Begley, S. (2001a) "Religion and the Brain", *Newsweek* (May), 14, 46–51.

Begley, S. (2001b) "We're Wired for Spirituality", *Newsweek* (May), 7, 52–57.

Bennet, M.R./Hacker, P.M.S. (2003) *Philosophical Foundations of Neuroscience*, Blackwell: Malden, USA/Oxford, UK.

Bergquist, L./Frangeur, R./Furuhagen, H./Gunneng, H./Westlund, B./Landtblom, A.M./Lindaräng, I./ Lindberg, E./ Rajamaa, R./Sjöberg, N./Witt-Brattström, E. (2003) *Birgitta – feminist, politiker och helgon: En antologi från Birgittajubileumsåret 2003,* Schuback, K. (ed.), Lindköping Universitet: Lindköping, Sweden.

Berk, L.E. (1998) *Child Development*, Allyn/Bacon: Boston, USA.

Bickle, J. (2003) *Philosophy and Neuroscience: A Ruthlessly Reductive Account*, Kluwer Academic Publishers: Norwell, USA.

Braud, W. (2002) "Brains, Science, Nonordinary & Transcendent Experiences", *NeuroTheology, Brain, Science, Spirituality, Religious Experiences*, Joseph, R. (ed.), University Press: California, USA, 123–135.

Brisman, B. (1987) *Mäster Eckehart, Medeltida Mystik*, Artos: Skellefteå, Sweden.

Brodd, S.E./Werdelin, L. (2004) "Are Models in Ecclesiology and Natural Sciences Structurally Homologous?", Paper presented to *Synpraxia Interdisciplinary Forum 2004*: Umeå, Sweden.

Brothers, L. (1997) *Friday's footprint: How Society Shapes the Human Mind*, Oxford University Press: Oxford, UK/New York, USA.

Brown, W.S. (2003a) "Neurophysiology", *Encyclopedia of Science and Religion*, van Huyssteen, W.J. (ed.), McMillan, W.J. Reference: New York, USA, 610.

Brown, W.S. (2003b) "Neuropsychology", *Encyclopedia of Science and Religion*, van Huyssteen, W.J. (ed.), McMillan, W.J. Reference: New York, USA, 610.

Brown, W.S. (2003c) "Neurosciences", *Encyclopedia of Science and Religion*, van Huyssteen, W.J. (ed.), McMillan, W.J. Reference: New York, USA, 610–617.

Brown, W.S./Murphy, N./Malory, N.H. (1998) *Whatever Happened to the Soul? Scientific and Theological Portraits of Human Nature*, Fortress Press: Minneapolis, USA.

Bruce, V./Green, P.R./Georgeson, M.A. (1997) *Visual Perception: Physiology, Psychology, and Ecology*, Psychology Press: Abingdon, UK.

Bråkenhielm, C.R. (1985) *Problems of Religious Experience*, Almqvist/Wiksell: Stockholm, Sweden.

Buck, R. (1980) *Änglar på Uppdrag*, Evangeliepress: Örebro, Sweden.

Cavero, M.J. (2004) "A Schematical View of the Ontologies Concept", *ECOOP workshop: Philosophy, Ontology and Information Systems*, University Leipzig: Leipzig, Germany.

Churchland, P.S. (1986) *Neurophilosophy*, The Massachusetts Institute of Technology: Massachusett, USA.

Churchland, P.S. (2002) *Brain-Wise, Neurophilosophy*, The Massachusetts Institute of Technology: Massachusett, USA.

Clayton, P. (1999) "Neuroscience, the Person and God: An Emergentist Account", *Neuroscience and the Person*, Russell, R.J./William, R./Stoeger, S.J/Ayala, F.J. (ed.), Vatican Observatory: California, USA, 181–214.

Cook, A. (1998) "Uncertainties of Science", *Science Meets Faith*, Watts, F. (ed.), Arrowsmiths: Bristol, UK.

Cools, L. (2001) "Wij staan erbuiten", *Hersenen, Bewustzijn, Zich op onszelf*, Oomen, P. (ed.), Valkhof Pers: Nijmegen, The Netherlands.

Cornwell, J. (1995) *Nature's Imagination, the Frontiers of Scientific Vision*, Oxford University Press: Oxford, UK.

Costello, J. (1989) "Psychosis or Religious Experience: Is there a difference?" *Guild Lectures*, No. 230.

Cowley, G./Kalb, C. (2003) "Our Bodies Our Fears", *Newsweek (*March), 3, 41–45.

Crick, F. (1995) *The Astonishing Hypothesis: The Scientific Search for the Soul*, Touchstone: New York, USA.

Davies, C.F. (1989) *The Evidential Force of Religious Experience*, Clarendon Press: Oxford, UK.

d'Aquili, E.G. (1978) "The Neurological Basis of Myth and Concepts of Diet" *Zygon*, vol. 13 (December), 257–275.

d'Aquili, E.G. (1982) "Senses of Reality in Science and Religion", *Zygon*, vol. 17 (December), 361–384.

d'Aquili, E.G. (1983) "The Myth-Ritual Complex: A Biogenetic Structural Analysis", *Zygon*, vol. 18 (September), 247–269.

d'Aquili, E.G. (1986) "Myth, Ritual and the Archetype Hypothesis: Does the Dance Generate the World", *Zygon*, vol. 21 (June), 141–160.

d'Aquili, E.G. (1993) "Apologia Pro Scriptura Sua, or Maybe we Got It Right After All" *Zygon*, vol. 28 (June), 152–165.

d'Aquili, E.G./Laughlin, C. (1975) "The biopsychological Determinants of Religious Ritual Behavior", *Zygon*, vol. 10 (March), 32–58.

d'Aquili, E.G./Laughlin, C./McManus, J. (1979) *The Spectrum of Ritual: A Biologenetic Structural Analysis*, Columbia University Press: New York, USA.

d'Aquili, E.G./Newberg, A.B. (1993a) "Liminality, Trance and Unitary States in Ritual and Meditation", *Studia Liturgia* 23:2–34.

d'Aquili, E.G./Newberg, A.B. (1993b) "Religious and Mystical States: A Neuropsychological Model", *Zygon*, vol. 28 (June), 177–199.

d'Aquili, E.G./Newberg, A.B. (1996) "Consciousness and the Machine", *Zygon*, vol. 31 (June), 235–252.

d'Aquili, E.G./Newberg, A.B. (1998a) "The Neuropsychology of Religion", *Science Meets Faith*, Watts, F. (ed.), Arrowsmiths: Bristol, UK, 73–91.

d'Aquili, E.G./Newberg, A.B. (1998b) "The Neuropsychological Basis of Religion: Or Why God Won't go Away", *Zygon*, vol. 33 (June), 187–201.

d'Aquili, E.G./Newberg, A.B. (1999) *The Mystical Mind: Probing the Biology of Religious Experience*, Fortress Press: Minneapolis, USA.

d'Aquili, E.G./Newberg, A.B. (2000) "The Neuropsychological of Aesthetic, Spiritual and Mystical States", *Zygon*, vol. 35 (March), 39–51.

Delio, I. (2003) "Brain Science and the Biology of Belief: a Theological Response", *Zygon*, vol. 38 (September), 573–585.

Dell, M.L. (1993) "Brain, Symbol & Experience: A Psychiatric and Theological Dialogue", *Zygon*, vol. 28 (June), 217–230.

Dennet, D.C. (1991) *Consciousness Explained*, Penguin Books/Clays: London, UK.

Dennet, D.C. (1996) *Kinds of Minds, Towards an Understanding of Consciousness*, Basic Schools: New York, USA.

Dupré, J. (2000) "Reductionism", *A Companion to the Philosophy of Science*, Newton-Smith, W.H. (ed.), Blackwell: Malden, USA/Oxford, UK/Melbourne, Australia/Berlin, Germany, 402–404.

Edwards, J. (1957/1973) *A careful and strict Enquiry into The modern prevailing notions of that Freedom of Will which, is supposed to be essential To Moral Agency, Virtue and Vice, Reward and Punishment, Praise and Blame*, Yale University Press: New Haven, USA/London, UK.

Edwards, J. (1959/2004) *A Treatise Concerning Religious Affections*, Electronic version, www.netlibrary.com.

Einhorn, S. (1998) *En Dold Gud, Om religion, vetenskap och att söka Gud*, Forum: Stockholm, Sweden.

Eliade, M. (1969) *The Quest, History and Meaning in Religion*, The University of Chicago Press: Chicago, USA/London, UK.

Feit, J.S. (2003) *Probing Neurotheology's Brain, or Critiquing on Emerging Quasi-Science*, a paper presented at the Annual Convention of the American Academy of Religion: Atlanta, USA.

Fenwick, P./Galliano, S./Coate, M.A./Rippere, V/Brown, D. (1985) "Psychic sensitivity, mystical experience, head injury and brain pathology", *British Journal of Medical Psychology*, 58, 35–44.

Geels, A. (2001) *Förvandlande ögonblick*, Norma: Skellefteå, Sweden.

Gellman, J. (2001) *Mystical Experience of God, A Philosophical Inquiry*, Ashgate: London, UK/Burlington, USA.

Gier, G.N. (1999) "Using Models to Represent Reality", *Models-Based Reasoning in Scientific Discovery*, Magnani, L./Nersessian, N. J./Thagard, P. (ed.), Kluwer/Phenum: New York, USA, 41–57.

Gleitman, H. (1994) *Psychology*, Norton, W.W./Co.: New York, USA.

Goldberg, E. (2001) *The Executive Brain, Frontal Lobes and the Civilized Mind*, Oxford University Press: Oxford, UK/New York, USA.

Granqvist, P./Fredrikson, K./Unge, P./Hagenfeldt, A./Valind, S./Larhammar, D./Larsson, M. (2005) "Sensed presence and mystical experiences are predicted by suggestibility not by the application of transcranial weak complex magnetic fields", *Neuroscience Letters*, vol. 379 (April), 1–6.

Gregersen, N.H./van Huyssteen, W.J. (1998) *Rethinking Theology and Science, Six Models for the Current Dialogue*, Eerdmans: Michigan, USA/Cambridge, UK.

Guttenplan, S. (1993) *A Companion of the Philosophy of Mind*, Blackwell: Oxford, UK.

Görman, U. (2002) "Modell, I. Naturwissenschaftlich", *Religion in Geschichte und Gegenwart*, Band 5, 1374–1375.

Haugh, J.F. (1995) *Science and Religion: From Conflict to Conversation*, Paulist Press: New York, USA.

Haugh, J.F. (2003) "Experience, Religious: Cognitive and Neurophysiological Aspects", *Encyclopedia of Science and Religion*, van Huyssteen, W.J. (ed.), Mcmillan Reference: New York, USA, 307–308.

Hick, J. (1989) *An Interpretation of Religion: Human Responses to the Transcendent*, Yale University Press: New Haven, USA/London, UK, 29–80.

Hick, J. (1996) "A Pluralist View", *Four Views on Salvation in a Pluralistic World*, Hick, J./ Pinnock, C.H./McGrath, A.E./Geivett, R.D./Phillips, W.G. (ed.), Zondervan: Michigan, USA.

Hobbs, J. (1993) *Religious Explanation and Scientific Ideology*, Peter Lang: New York, USA.

Hof, H. (2000) *Bli mer människa*, Benedictum: Uppsala, Sweden.

Holmes, H.R. (1993) "Thinking about Religion and Experiencing the Brain: Eugene d'Aquili's Biogenetic Structural Theory of Absolute Unitary Being", *Zygon*, vol. 28 (June), 102–115.

Holte, R. (1984) *Människa, Livstolkning, Gudstro*: Teorier och metoder inom tros- och livsåskådningsvetenskapen, Doxa: Bodafors, Sweden.

Häring, H. (2001) "Het binnenste kwijtgeraakt? Mensen als beeld van God", *Hersenen, Bewustzijn, Zich op onszelf*, Oomen, P. (ed.), Valkhof Pers: Nijmegen, The Netherlands.

Ingvar, D.H. (2001) *Hjärnans futurum, Cerebrala essäer om medvetande, minne, språk, musik, humör, vilja, drömmar, känslor och koma*, Atlantis: Stockholm, Sweden.

James, W. (1997) *The Varieties of Religious Experience*, Touchstone: New York, USA.

James, W. (2002a) "The Reality of the Unseen", *NeuroTheology, Brain, Science, Spirituality, Religious Experiences*, Joseph, R (ed.), University Press: California, USA, 135–143.

James, W. (2002b) "Conversion and Religious Experience", *NeuroTheology, Brain, Science, Spirituality, Religious Experiences*, Joseph, R. (ed.), University Press: California, USA, 203–227.

James, W. (2002c) "Mysticism", *NeuroTheology, Brain, Science, Spirituality, Religious Experiences*, Joseph, R. (ed.), University Press: California, USA, 227–241.

Jeffner, A. (1975) *Filosofisk religionsdebatt*, Tryckmans: Stockholm, Sweden.

Joseph, R. (2001) "The Limbic System and the Soul: Evolution and the Neuroanatomy of Religious Experience", *Zygon*, vol. 36 (March), 105–136.

Jung, C.G. (1958) "Psychology and Religion: West and East", *The Collected Works of C.G. Jung*, Sir Read, H./Fordham, M./Adler, G./McGuire, W. (ed.), Routledge/Kegan P.: London, UK. vol. II.

Kalb, C. (2003) "Faith & Health", *Newsweek* (November), 10.

Kephart, B. (2001) "Medical alumnus and faculty member Dr. Andrew Newberg probes deep inside the brain in a quest to understand faith", *The Gazette: Looking for God* (May/June), www.upenn.edu/gazette/0501/kephart.htm.

King James Version (1984) *The Holy Bible,* Standard Text Edition, Cambridge University Press: Cambridge, UK.

Kim, J. (1992) "Downward Causation in Emergentism and Non-reductive Physicalism", *Emergence or Reduction? Essays on the Prospects of Nonreductive Physicalism*, Beckmann, A./ Flohr, H./Kim, J. (ed.), Walter de Gruyter: New York, USA, 119–138.

Klunk, W. (2003) "New Imaging Agent Detects Amyloid Plagues in Alzheimer Disease", *Neurology Today,* 3 (12):1, 15–16.

Kosslyn, S.M. (2003) "The Wet-wiring of Human Potential", *News & Opportunities, Research in Science and Theology*, vol. 3 (Januray), 6/25.

Kozard, M. (1989) "Religious Experience Was Not Correctly Defined", *Journal of Neuropsychiatry*, 10, 4:475.

Lagercrantz, H. (2002) *Hjärnan och medvetandet*, Nya Doxa: Nora, Sweden.

Landtblom, A.M. (2000) "Hade heliga Birgitta epilepsi?", *Läkartidningen,* vol. 97, no. 51–52, 6044–6048.

Landreth, A. (2003) "Brain-Wise: Studies in Neurophilosopy by Patricia Smith Churchland, *The Human Nature Review*, MITT Press: Califronia, USA, vol. 3, 455458.

LaPiere, R. (1938) "The Sociological Significance of Measurable Attitudes", *American Sociological Review,* vol. 3, nr. 2 (April), 175-182.

Laughlin, D.Jr./McManus, J./d'Aquili, E.G. (1993) "Mature Contemplation" *Zygon* vol. 28 (June), 133–175.

Ledoux, J.E./Hirst, W. (1986) *Mind and Brain: Dialogues in Cognitive Neuroscience*, Cambridge University Press: Cambridge, USA.

Ledoux, J.E. (1999a) "Emotions-A view through the Brain", *Neuroscience and the Person*, Russell, R.J./Murphy, N./Meyering, T.C./Arbib, M.A. (ed.), Vatican Observatory: California, USA, 101–118.

Ledoux, J.E. (1999b) "How I've looked for Them in the Brain", *Neuroscience and the Person*, Russell, R.J./Murphy, N./Meyering, T.C./Arbib, M.A. (ed.), Vatican Observatory: California, USA, 41–44.

Ledoux, J.E. (2002) *Synaptic Self*, Penguin Group: London, UK.

Liebert, R.M./Liebert, L.L. (1995) *Science and Behavior, an introduction to methods of psychological research*, Prentice-Hall: New Jersey, USA.

Lockwood, M. (1989) *The Mind, the Brain and the Quantum*, Blackwell: London, UK.

Marcel, A.J./Bisiach, E., 1998, *Consciousness in Contemporary Science*, Oxford Science Publications: Oxford, UK.

McCulloch, G. (1995) *The mind and Its World*, Routledge: London, UK/New York, USA.

MacLennan, B. (2002) "Evolutionary Neurotheology & the Varieties of Religious Experience", *NeuroTheology, Brain, Science, Spirituality, Religious Experiences*, Joseph, R. (ed.), University Press: California, USA, 305–315.

Matlin, M.W./Foley, H.J. (1997) *Sensation and Perception*, Allyn/Bacon: Boston, USA.

Meijsing, M. (2001) "Het probleem van het bewustzijn", *Hersenen, Bewustzijn, Zich op onszelf*, Oomen, P. (ed.), Valkhof Pers: Nijmegen, The Netherlands.

Mühling-Schlapkohl, M. (2002) "Modell, II. Religionsphilosophisch", *Religion in Geschichte und Gegenwart*, Band 5, 1375.

Mumenthaler, M./Mattle, H. (2004) *Neurology*, Georg Thieme: New York, USA.

Murphy, N. (1996) *Beyond Liberalism and Fundamentalism: How Modern and Postmodern Philosophy Set the Theological Agenda*, Trinity Press: Valley Forge, USA.

Murphy, N. (1998) "Supervenience and the Nonreducibility of Ethics to Biology" *Evolutionary and Molecular Biology: Scientific Perspectives on Divine Action*, Russell, R.J./William, R./Stoeger, S.J/ Ayala, F.J. (ed.), Vatican Observatory: California, USA, 463–489.

Newberg, A.B. (2001) "Putting the Mystical Mind Together", *Zygon*, vol. 36 (September), 501–507.

Newberg, A.B. (2002) "Bringing 'Neuro' and 'Theology' Together", *NeuroTheology, Brain, Science, Spirituality, Religious Experiences*, Joseph, R. (ed.), University Press: California, USA, 145–147.

Newberg, A.B. (2003) "Exobiology", *Encyclopedia of Science and Religion*, van Huyssteen, W.J. (ed.), Mcmillan Reference: New York, USA, 306–311.

Newberg, A.B./Alavi, A./Baime, M./Mozley, P.D./d'Aquili, E.G. (1997) "The Measurement of Cerebral Blood Flow during the Complex Cognitive Task of Meditation using HMPAO-Spect Imaging", *Journal of the Neurological Medicine*, vol. 38, No 5, 95P.

Newberg, A.B./Alavi, A./Baime, M./Pourdehnad, M./Santanna, J./d'Aquili, E.G. (2001a) "The Measurement of reginal Cerebral Blood Flow during the Complex Cognitive Task of Meditation: a Preliminary SPECT Stud", *Psychiatry Research Neuroimaging Section*, 106, 113–122.

Newberg, A.B./d'Aquili, E.G. (2000) "The Creative Brain/The Creative Mind", *Zygon,* vol. 35 (March), 53–65.

Newberg, A.B./d'Aquili, E.G./Rause, V. (2001b) *Why God Won't go Away*, Ballantine: New York, USA.

Newberg, A.B. (2004) webpage: www.andynewberg.com.

Newton-Smith, W.H. (2000) "Explanation", A Companion to the Philosophy of Science, Blackwell: Malden, USA/Oxford, UK/Melbourne, Australia/Berlin, Germany, 127–133.

Nijs, P./Meyers, K. (2001) *De onbereikbaarheid van de geliefde, pareltjes van Nederlandse en Rijnlandse Mystiek*, Peeters: Leuven, Belgium.

Oomen, P. (ed.) (2001) *Hersenen, Bewustzijn, Zich op onszelf*, Valkhof Pers: Nijmegen, The Netherlands.

Oomen, P. (2003) "Neurotheology", *Encyclopedia of Science and Religion*, van Huyssteen, W.J. (ed.), Mcmillan Reference: New York, USA, 617–618.

Otto, R. (1925) *Das Heilige: Uber das Irrationale in der Idee des Göttlichen und sein Verhältnis zum Rationalen*, Klotz, L.: Gotha, Germany.

Otto, R. (1970) *The Idea of the Holy*, Oxford University Press: London, UK.

Persinger, M.A. (1984) "Striking EEG Profiles from Single Episodes of Glossolalia and Transcendental Meditation", *Perceptual and Motor Skills*, 76, 127–133.

Persinger, M.A. (1985) "Temporal Lobe Signs: Electroencephalographic Validity and Enhanced Scores in Special Populations", *Perceptual and Motor Skills*, 60, 831–842.

Persinger, M.A. (1987) *Neuropsychological Bases of God Beliefs*, Praeger: New York, USA.

Persinger, M.A. (1991) "Preadolescent Religious Experience Enhances Temporal Lobe Signs in Normal Young Adults", *Perceptual and Motor Skills*, 72, 453–454.

Persinger, M.A. (1992) "Right Hemisphericity, Low Self-Esteem, and Unusual Experiences: A Response to Vingiano", *Perceptual and Motor Skills*, 75, 568–570.

Persinger, M.A. (1993a) "Paranormal and Religious Beliefs may be Mediated Differentially by Subcortical and Cortical Phenomenological Processes of the Temporal (Limbic) Lobes", *Perceptual and Motor Skills,* 76, 247–251.

Persinger, M.A. (1993b) "Transcendental Meditation TM and General Meditation are Associated with Enhanced Complex Partial Epileptic-like Signs: Evidence for 'Cognitive' Kindling?" *Perceptual and Motor Skills*, 76, 80–82.

Persinger, M.A. (1993c) "Vectorial Cerebral Hemisphericity as Differential Sources for the Sensed Presence, Mystical Experiences and Religious Conversions", *Perceptual and Motor Skill,* 76, 915–930.

Persinger, M.A. (1997) "'I would kill in God's name': role of sex, weekly church attendance, report of a religious experience, and limbic lavility", *Perceptual and Motor Skills*, Aug. 95, 128–130.

Persinger, M.A. (2002a) "Experimental Simulation of the God Experience", *NeuroTheology, Brain, Science, Spirituality, Religious Experiences*, Joseph, R. (ed.), University Press: California, USA, 279–293.

Persinger, M.A. (2002b) "Neurobehavioral and Neurometabolic (SPECT) Correlates of Paranormal Information involvement of the Right Hemisphere and its Sensitivity to Weak Complex Magnetic Fields", *International Journal of Neuroscience*, 112:197–224.

Persinger, M.A. (2002c) "The Temporal Lobe: The Biological Basis of the God Experience", *NeuroTheology, Brain, Science, Spirituality, Religious Experiences*, Joseph, R. (ed.), University Press: California, USA, 273–279.

Persinger, M.A. (2003) "The Sensed Presence within Experimental Settings: Implications for the Male and Female Concept of Self", *The Journal of Psychology*, 137 (1) 5–16.

Persinger, M.A. (2004) webpage, www.laurentian.ca.

Peterson, G.R. (2002) "Mysterium Tremendum", *Zygon*, vol. 37 (June), 237–254.

Peterson, G.R. (2003) *Minding God, Theology and the Cognitive Sciences*, Fortress Press: Minneapolis, USA.

Peterson, M./Hasker, W./Reichenbach, B./Basinger, D. (1996) *Philosophy of Religion: Selected Readings*, Oxford University Press: Oxford, UK.

Peterson, M./Hasker, W./Reichenbach, B./Basinger, D. (1999) *Religionsfilosofiska texter*, Nya Doxa: Nora, Sweden.

Peterson, M./Hasker, W./Reichenbach, B./Basinger, D. (2003) *Reason & Religious Belief, an Introduction to the Philosophy of Religion*, Oxford University Press: Oxford, UK.

Philipchalk, R./Mueller, D. (2000) "Glossolalia and Temperature Change in the Right and Left Cerebral Hemispheres", *The International Journal for the Psychology of Religion*, 10 (3) 181–185.

Phillips, D.Z. (1976) *Religion without explanation*, Blackwell: Oxford, UK.

Pols, E. (1999) "Knowing God directly", *International Journal for Philosophy of Religion*, 45: 31–49.

Popper, K.R./Eccles, J.C. (1981) *The Self and Its Brain, An argument for Interactionism*, Springer: Berlin /New York/London.

Proudfoot, W. (1985) *Religious Experience*, University of California Press: Berkley, USA/ London, UK.

Proust, M. (1913) "Remembrance of Things Past." *Swann's Way: Within a Budding Grove*, vol. 1, The definitive French Pleiade edition translated by Moncrieff, C.K.S./Kilmartin, T., Vintage: New York, USA, 48–51. www.haverford.edu/psych/ddavis/p109g/proust.html, spring (2004).

Quinn, P.L./Taliaferro, C. (1999) *A Companion to Philosophy of Religion*, Blackwell: Malden, USA/Oxford, UK/Melbourne, Australia/Berlin, Germany.

Ramachandran, V.S. (2003) *The Emerging Mind*, The Reich Lectures, Profile Books: London, UK

Ravous, J.P. (2000) *l'Unité des sciences: Expliquer la nature et comprendre l' home*, Le Pommier-Fayard: Turin, France.

Runehov, A.L.C. (2004) "A Neuro-psychological Explanation of Religious Experience?", *Studies in Science and Theology*, Görman, U./Drees, W.B./Meisinger, H. (ed.), vol. 9, 59–68.

Runehov, A.L.C. (2006) "A Being or To Be? Philosophical Thoughts about Future Research on Neuroscience and Religions and the Need for Interdisciplinarity", *European Journal of Science and Theology,* vol. 2, No. 1 (March), 55–66.

Russel, R.J./Wegter-McNelly, K. (2003) "Science and Religion", *Encyclopedia of Science and Religion*, van Huyssteen, W.J. (ed.), Mcmillan Reference: New York, USA, 746–748.

Salzman, M. (2000) *Lying Awake*, Knopf, A.A.: New York, USA.

Saver, J.L./Rabin, J. (1997) "The Neural Substrates of Religious Experience", *Journal of Neuropsychiatry and Clinical Neuroscience*, vol. 9, no. 3, 498–510.

Schleiermacher, F. (1923) *Tal över Religionen*, Bonniers: Stockholm, Sweden.

Schleiermacher, F. (1928)/(1976) *The Christian Faith*, Clark, T/T: Edingburgh, UK.

Schopenhauer, A. (1992) *Värden som vilja och föreställning*, Nya Doxa: Nora, Sweden.

Scott, A. (1995) *Stairways to the Mind*, Copernicus: New York, USA.

Searl, J.R. (1992) *The Rediscovery of the Mind*, The MIT press: Cambridge, UK

Silberstein, M. (2002) "Reduction, Emergence and Explanation", *The Blackwell Guide to the Philosophy of Science*, Machamer, P./Silberstein, M. (ed.), Blackwell: Oxford, UK, 80–107.

Sinkinson, C. (2001) *The Universe of Faiths: A Critical Study of John Hick's Religious Pluralism*, Paternoster Press: Carlisle, UK/Waynesboro, USA.

Spezio, M.L. (2001) "Understanding Biology in Religious Experience: The Biogenetic Structuralist Approach of Eugene d'Aquili and Andrew Newberg", *Zygon,* vol. 36 (September), 477–484.

Stenmark, M. (1996) "Villkor för en godtagbar teologisk och filosofisk forskning", *Svensk Teologisk Kvartalskrift*, Årg. 72.

Stenmark, M. (2001b) *Scientism: Science, Ethics and Religion*, Ashgate: Aldershot, UK/Burlington, USA.

Stenmark, M. (2004) *How to Relate Science and Religion*, Eerdmans: Michigan, USA/Cambridge, UK.

Swaad, D. (2001) "Hersenen, bewustzijn en geloof: neurobiologische aspecten", *Hersenen, Bewustzijn, Zich op onszelf*, Oomen, P. (ed.), Valkhof Pers: Nijmegen, The Netherlands.

Swinburne, R. (1979) *The Existence of God*, Clarendon: Oxford, UK.

Taliaferro, C. (1994) *Consciousness and the Mind of God*, Cambridge University Press: Cambridge, UK.

Teller, P. (1992) "A Contemporary Look at Emergence", *Emergence or Reduction? Essays on the Prospects of Nonreductive Physicalism*, Beckmann, A./Flohr, H./Kim, J. (ed.), Walter de Gruyter: New York, USA, 139–156.

The Medical Advisory Secretariat (2004) *Repetitive Transcranial Magnetic Stimulation for the Treatment of Major Depressive Disorder*, Health Technology Scientific Literature and Policy Review, Ministry of Health and Longterm Care: Toronto, Canada.

The Royal Australian and New Zealand College of Psychiatrists (2003) "Transcranial Magnetic Stimulation (TMS)", *Positions Statement* No. 40 (June), GC2003/2, R61.

Thiselton, A.C. (2002) *A Concise Encyclopedia of the Philosophy of Religion*, Oneworld: Oxford, UK.

Trillingsgaard, A./Dalby, M./Ostergaard, J. (1999) *Barn som är annorlunda*, Studentliteratur: Lund, Sweden.

Tye, M. (1999) "Phenomenal Consciousness: The Explanatory Gap as a Cognitive Illusion", *Mind*, vol. 108, 705–725.

van Huyssteen, W.J./Howell, N.R./Gregersen, N.H./Wildman, W.J./Barbour, I.G./Valentine, R. (2003) *Encyclopedia of Science and Religion*, Mcmillan Reference: New York, USA.

Waksh, V./Cowey, A. (2000) "Transcranial magnetic stimulation and cognitive neuroscience", *Nature Reviews, Neuroscience*, vol. I (October), 73–79.

Wildman, W.J./Brothers, L.A. (1999) "A Neuropsychological-Semiotic Model of Religious Experiences", *Neuroscience and the Person*, Russell, R.J./Murphy, N./Meyering, T.C./Arbib, M.A. (ed.), Vatican Observatory: California, USA, 347–416.

Williams, P.S. (2002) *The Case for Angels*, Paternoster Press: Carlisle, UK/Waynesboro, USA.

Watts, F. (1999) "Cognitive Neuroscience and Religious Consciousness", *Neuroscience and the Person*, Russell, R.J./William, R./Stoeger, S.J/Ayala, F.J. (ed.), Vatican Observatory: California, USA, 327–346.

Watts, F. (2002a) "Interacting Cognitive Subsystems & Religious Meanings", *NeuroTheology, Brain, Science, Spirituality, Religious Experiences*, Joseph, R. (ed.), University Press: California, USA, 183–189.

Watts, F. (2002b) *Theology and Psychology*, Ashgate: Aldershot, UK/Burlington, USA.

Watts, F. (2003) "Brain, Mind and Soul", *Science Meets Faith*, Arrowsmiths: Bristol, UK, 59–72.

Wiebe, P.H. (1997) *Visions of Jesus*, Oxford University Press: Oxford, UK/New York, USA.

Wolkowski, Z.W. (1993) "The Concept of Coherence Considered as a Systems Isomorphism", *Systems Research and Behavioral Science,* vol. 9, issue 4, 61–66.

Woodward, J. (2002) "Explanation", *The Blackwell Guide to the Philosophy of Science*, Machamer, P./Silberstein, M. (ed.), Blackwell: Oxford, UK, 37–54.

Yandell, K. (1993) *The Epistemology of Religious Experience*, Cambridge University Press: Cambridge, UK.